STEVIA REBAUDIANA

CHEMICAL COMPOSITION, USES AND HEALTH PROMOTING ASPECTS

FOOD AND BEVERAGE CONSUMPTION AND HEALTH

STEVIA REBAUDIANA

CHEMICAL COMPOSITION, USES AND HEALTH PROMOTING ASPECTS

DAVID BETANCUR-ANCONA

AND

MAIRA SEGURA-CAMPOS

EDITORS

NOTICE TO THE READER

Library of Congress Cataloging-in-Publication Data

ISBN: 978-1-63463-335-2

Library of Congress Control Number: 2014953193

Published by Nova Science Publishers, Inc. † New York

CONTENTS

PREFACE

Human health is greatly endangered and characteristic diseases developed in the last decade due to excessive intake of "harmful" sugars present in wide range of food products. From a medical point of view, increased use of products enriched with sugar favor the development of chronic diseases. Obesity and diabetes are ones of the major diseases of modern mankind and from the health aspect a great emphasis in the prevention thereof was placed. Increased interest of the consumer to reduce sugar intake through the food leads to higher popularization of products that instead of sucrose contain artificial sweeteners. Most artificial sweeteners with a high degree of sweetness are produced from synthetic ingredients or derived exclusively by chemical synthesis in the laboratory. Modern medical researches show that most artificial sweeteners are harmful for human health. Therefore, more and more sweeteners extracted from natural materials are used.

Stevia rebaudiana is a small perennial shrub that belongs to the family aster or chrysanthemum family. It grows primarily in the Amambay mountain range of Paraguay. However over 200 species of stevia have been found around the world. *Stevia rebaudiana* is the only species at present, which possesses the ability to sweeten. Stevioside is natural sweetener isolated from the leaves of plant *Stevia rebaudiana* and it is up to 300 times sweeter than sucrose. Dry leaves of stevia are sweeter approximately 10 to 15 times than sucrose while glycemic index is zero, so it is sweetener with no caloric value and with proven non-toxic effect on human health. The leaves of stevia naturally contain a mixture of 8 sweet diterpene glycosides: stevioside, steviolbioside, rebaudiosides (A, B, C, D and E) and dulcoside A. Besides the high stevioside content, stevia plant is characterized as a good source of protein, dietary fiber, minerals and essential amino acids. *S. rebaudiana* has attracted economic and scientific interests due to the sweetness and the supposed therapeutic properties of its leaf.

Stevia and stevioside have been applied as substitutes for saccharose, for treatment of diabetes mellitus, obesity, hypertension and caries prevention, and a number of studies have suggested that, besides sweetness, stevioside, along with related compounds which include rebaudioside A, steviol and isosteviol, may also offer therapeutic benefits, as they have anti-hyperglycemic, antihypertensive, anti-inflammatory, anti-tumour, anti-diarrhoeal, diuretic, and immunomodulatory effects. The leaves of Stevia has functional and sensory properties superior to those of many other high-potency sweeteners, and is likely to become a major source of high-potency sweetener for the growing natural food market in the future.

The purpose of this book is to provide an insight into the possibilities of applying stevia as a sweetener for commercial purposes; biochemical and nutritional content of the stevia leaves and use of stevia leaves as a basic raw material for the various stevia products. Emphasis will be placed on the remarkable potential of stevia as an intense high-potency sweetener together with its functional and health-promoting properties, making thereby a contribution in enhancing the importance of *S. rebaudiana* as a promising new agricultural crop.

The editors are grateful to all experts and renowned authors for their state-of-the-art compilation of recent rapid developments in this field, which made possible the publication of this book. We thank members of the production team at NOVA for their time, effort, advice and expertise. We believe that this book deserves a broad readership in the disciplines of food science, nutrition pharmaceuticals, biochemistry and functional foods. This book could also be useful as a comprehensive reference book by senior undergraduate and graduate students, as well as by agricultural and food industries.

Maira Rubi Segura Campos
David Betancur Ancona
Universidad Autónoma de Yucatán, Mexico

In: Stevia rebaudiana
Editors: D. Betancur and M. Segura

ISBN: 978-1-63463-335-2
© 2015 Nova Science Publishers, Inc.

Chapter 1

STEVIA REBAUDIANA BERTONI, A CROP WITH A PRODUCTIVE, SWEETENER AND MEDICINAL POTENTIAL IN MEXICO

Genovevo Ramírez Jaramillo[1,•]
and Y. B. Moguel-Ordoñez[2,*]

[1]Researcher of Center for Development Cooperation of the Tropics
(CECODET) CIRSE-INIFAP, Cologne Diaz Ordaz,
Merida, Yucatan, Mexico
[2]Researcher of Mocochá Experimental Field, Old Road Merida-Motul,
Mococha, Yucatan, Mexico

ABSTRACT

The increase in demand for Stevia in the domestic and international market and the lack of raw material availability has led to initiatives to increase the area of cultivation by the government, industry, private business and the social sector in Mexico. To determine the production potential of *Stevia rebaudiana* Bertoni, databases of the Mexico climate, soil and digital elevation and geographic information systems were used to process information and generate maps. To define the sweetening potential an analytical technique was used, wherein the sheet content of two of its major glycosides (Stevioside and Rebaudioside-A) and their

• E-mail: ramirez.genovevo@inifap.gob.mx, Phone (999) 1 96 11 83 Ext 601.
* E-mail: moguel.yolanda@inifap.gob.mx, Phone: (991) 9 16 22 15 Ext. 143.

place in the health sector of Mexico was determined. A descriptive analysis of the main problems of chronic non-communicable diseases in Mexico and themes of research and development were aimed at addressing these problems. Based on the methodology used, in Mexico there are more than 3 million hectares produced in optimum conditions and more than 2 million suboptimal, so it is feasible to increase production especially in the Pacific region, the Gulf and to a lesser extent in the Yucatan Peninsula. The sweetening power in the Morita II material is where the highest contents of Rebaudioside-A were identified, being in the range of 6.83 to 13.31, while in creole materials its range was 1.72 to 2.93. The potential medicinal uses of this crop for non-transmissible chronic diseases such as obesity, diabetes, hypertension and even cancer are very broad but more research needs to be done.

INTRODUCTION

The Stevia (*Stevia rebaudiana* Bertoni) plant is originally from the Amambay mountain range located between the south of Brazil and the northeast of Paraguay, where the Guaraní Indigenous people have used the plant since ancestors' time. The crop is known in their dialect as Ka'a He'ê or "hierba dulce" (Sweet herb).

Stevia is currently attracting attention from various areas of the population as an important sweetener, and a possible substitute for artificial sweeteners, due to the fact that the glycosides that are extracted from the dried leaves are 200 to 300 times sweeter than sucrose. Even though Stevia is a powerful sweetener, researchers are considering that its greatest contribution might be in the health sector.

Demand for natural sweeteners has increased worldwide, mainly due to the side effects produced by artificial or synthetic sweeteners. Japan is consuming *S. rebaudiana,* not only as a sweetener but also as a food additive. Japan began using Stevia after World War II because of the difficulties faced in obtaining sugar and its overpricing in the international market. This plant is being used regularly by China, and it is completely legal. In South American countries, it is customarily sold in supermarkets. In North America and the European Union it was introduced from 2008 to 2011 as a natural sweetener.

In Japan in 1998, Stevia Laboratory Ltd. emerged with an initial investment of 30.000.000 ¥. The main goal of this organization was to research different applications of *S. rebaudiana* as a sweetener, for health issues, agricultural, fishing and livestock related use, as well as cosmetic use.

By 1997, the lab had 25 patents plus others in the process of approval (JBB 2014).

70% of the world production of Stevia is primarily used to extract crystals, mainly from the glycosides in the leaves: Stevia and Rebaudioside-A, while the remaining 30% is used in herbal medicine (Shanks et al., 2004).

S. rebaudiana can be an innovative and profitable crop, showing promising conditions for national and international markets. The use of stevia, as an herb, or as an industrialized product derived from this crop is presented as an alternative to the use of artificial sweeteners such as aspartame, saccharin, cyclamate, etc., products that have been questionable for the health risk they may represent to the consumers that are for the most part, diabetics, obese or weight loss seeking individuals. To respond to the interest of producers, investors and marketers as well as the increased farming in Mexico, there is a need to identify the potential regions for *S. rebaudiana* production, its plant quality as a sweetener and its possibilities to improve some of the health issues that affect the Mexican population.

METHODOLOGY

Databases were used to determine the production potential of Stevia in Mexico, with information about climate, soil and digital elevation. Information systems were also used to process the databases and generate maps: to define its sweetener potential; an analytical technique in which the content of the main glycosides were analyzed (Stevioside and Rebaudioside-A) to determine the contents in one stevia leaf. Studies on stevia and its possible contributions to the health sector in Mexico, as well as other research subjects are necessary.

The Productive Potential of *S. rebaudiana* Bertoni

The crop distribution worldwide is marginalized by climatic limits, by defects or excess in vital needs for individuals that are part of different biotypes. Even during sowing processes, plants are under asynchronous variations from elements in climate; this is the most determining factor of success probabilities in a crop (Baradas, 1994., Benacchio, 1982., Doorenbos and Kassam, 1979., FAO, 1993., Garcia, 1988., Ruiz et al., 1999., Warrington and Kanemasu, 1983).

Based on the previous information, potential areas were determined, considering the areas open for farming, therefore, the expansion potential is much larger in the country (INIFAP, 1993a; INIFAP, 1993b; INIFAP, 1993c; Ramírez, 1995; Ramírez, 200; Ramírez, 2006). Optimal and suboptimal conditions were also taken into account for the development of the crop, the elevation, altitude, climate, soil, temperature, and precipitation variables (Table 1).

The analysis and information process was done with ArcView version 3.3 software. This is a program developed by ESRI. Through this system of geographically referenced data, we can analyze characteristics and distribution patterns of this data with vectorial maps and generate results with the information obtained (ESRI, 1996).

In the processing of information, Arc View generates several different forms of result reports, such as Views, Tables, Graphics, layouts and scripts.

All information processed through Arc View is named and saved in an ASCII format and will always have the extension *.apr.

Table 1. Agroecological Requirements *Stevia rebaudiana* Bertoni

Parameter	Optimum	Suboptimal	Unsuitable
Climate	Tropical and Subtropical	Tempered	Cold
Temperature	18-30	15-18 30-43	Under 15 Over 43
Precipitation	1000 -1400	500 – 1000 1400-2000	Under 500 Over 2000
Altitude	0 - 500	500 1200	Over 1200
Soil Depth. Drainage Soil Texture Soil Type	20 50 cm Groad Loam and Sandy Loam Luvisols, Nitisols, Regosols y Fluvisols	10 20 cm Regul Heavy Cambisols Rendzinas (10 %)	Under 10 Low Heavy Soloncha, Vertisoles, Gleysoles Litosoles
pH	5 7.0	4 4.99 7.1 7.5	Under 4 Over 7.5

Source: Ramírez et al, 2012.

Crop requirements were identified and analyzed for technical characteristics that are appropriate for the development of *S. rebaudiana,* for each variable considered in the study.

Considering the conditions as: optimal (high), suboptimal (medium), and not suitable, further on the potential regions for farming were determined by cartographic intersections in Arc View 3.3. In the vector process, maps are generated through cartographic intersections inserted among polygons, and the potential levels are described and maintained in each process of intersection. In this way the final map presents the information of all the variables that were intercepted. These maps are more representative than raster models; vectorial maps are more exact when generating surface calculations since they are polygonal maps (Figure 1).

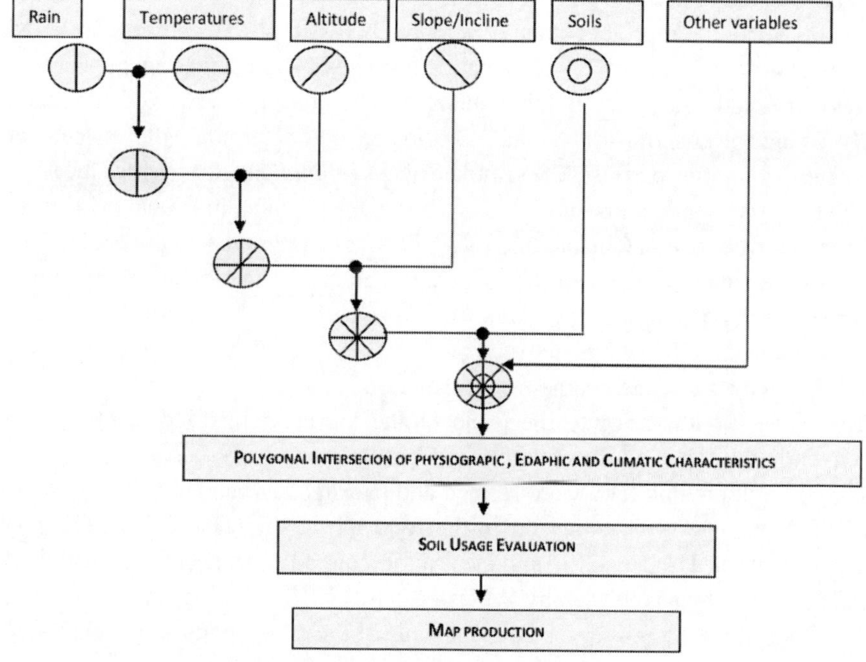

Figure 1. Intersection Process of Vectorial Maps to Identify Potential Classes.

Sweetener Potential

The highest content of glycosides in *S. rebaudiana* were found in the leaves, either fresh or dried, and they are up to 15 times sweeter than sugar. The contents of glycosides found in the Steviol crystals are up to 300 times higher than sugar cane.

S. rebaudiana leaves have several glycoside compounds. These compounds give the sweet taste. All glycosides in Stevia share the same molecular nucleus: Steviol. The only difference among several glycosides in stevia is the amount and layout of the glucose molecules adhered to the nucleus.

Based on the weight of dried up ground leaves of *S. rebaudiana*, the four main Glycosides are Dulcoside-A with 0.3%, Rebaudioside-C at 0.6%, Rebaudioside-A at 3.8% and Stevioside at 9.1 % (Brandle et al., 1998).

In Mexico, a project was developed between 2010 and 2011, in 15 farming research and technology validation sites. The adaptation of the crop was observed, along with the content in the leaves of two of the main glycosides, Rebaudioside-A and Stevioside to determine the sweetener potency. For this purpose 15 samples of *S. rebaudiana* leaves produced in states in the south and southeast of the country, eight in Yucatan, five in Quintana Roo, one in Chiapas and one in Veracruz (Ramírez et al., 2011).

In the states of Yucatan, Chiapas and Veracruz, samples of the Morita II variety were used, and in Quintana Roo, besides Morita II, samples of Criollo of Paraguay and SM_1 were analyzed.

The leaves obtained came from plantations that had been established for at least three months prior to the project. The leaves were dried at 60°C in a conventional stove or in sunlight, and ground until they became a powder using a cyclonic mil; they were labeled and packaged as needed.

The samples were analyzed in the food lab in the chemical engineering department of UADY (Universidad Autónoma de Yucatán), the analytical technique used was reported by Woelwer *et al.*, 2010, showing modification in the dilution of the samples, since the signals of the compounds was higher than the standard curve in stevioside and Rebaudioside-A.

A design chosen at random was used to evaluate differences between Rebaudioside-A and Stevioside in the leaves of the varieties utilized, as well as states and sites for farming. A variance and comparison average was analyzed by the Fisher Method (LSD), using the Stat Graphics Plus version 5.1 package.

Medicinal Potential of *S. rebaudiana*

Besides the sweetener properties of *S. rebaudiana*, it also provides health benefits, such as the control of blood sugar levels, some vasodilators and antibiotic properties. At the present time, research has multiplied in Japan, the US, Sweden, Italy and Germany. Just in Japan there are 1500 scientific studies that support the consumption and use of *S. rebaudiana*. There are also, a large number of studies in Denmark, Brazil, Israel, Canada, etc. (www.alimentacion-sana.org)

The importance of *S. rebaudiana* lies within its natural properties and the benefits it brings when consumed. Surprising results can be observed in reducing obesity, diabetes, and prevention of certain types of pain (Fischer, 2013).

In México, there are a few scientific studies of the use of *S. rebaudiana* related to medical aspects. However, there are several advances in other countries that are significant references for consulting and promoting interest in the health sector, which may encourage studies in this direction to search for alternatives and solutions to problems that have dramatically increased in the country, such as obesity, diabetes, hypertension, and cancer, among others.

A documented research was developed to locate the main studies about *S. rebaudiana* focused on health. The documented investigation is characterized by the use of information that collects, analyses, and presents coherent results, and uses research procedures such as analysis, synthesis, deduction and induction, etc.

RESULTS AND DISCUSSION

Productive Potential

In the tropical region stevia has a wide range of adaptation, from 0 to 200 meters above sea level, but it is in subtropical climates where leaves with the best quality are obtained. The distribution of agro ecological conditions for growing Stevia in Mexico is presented in the following maps:

Altitude

From the biological point of view, altitude affects the growth of plants; the length of internodes, the size of leaves and the glycoside content, since altitudes more than 1,200 meters above sea level reduces their accumulation.

The range of height above sea level that optimally adapts and develops Stevia is between 0 to 500 meters above sea level, it is considered suboptimal between 500 and 1200 and above 1200 not suitable (Figure 2).

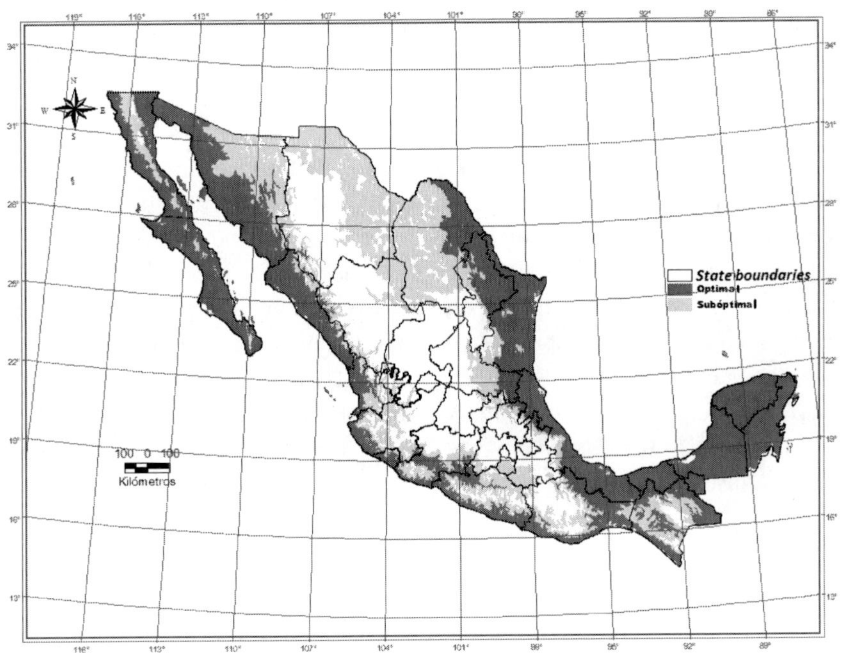

Figure 2. Altitude Optimal and Suboptimal in Mexico.

Precipitation

The demand for humidity for the growth of this crop is high and continuous, that is to say it should never lack water during different stages of development.

The natural distribution of this crop is observed in areas where the average annual precipitation is high (1.400 mm to 1.600 mm) and usually uniform between 100 to 120 mm per month.

In Mexico, even if it is feasible to locate areas with rainfall within the optimum range, the need for supplemental irrigation rises. Because in most of Mexico there is a 4 to 6 month drought period, which the crop does not tolerate, therefore the use of irrigation systems is necessary (Casaccia and Alvarez, 2006).

Optimum condition was considered in the range of 1,000 to 1,400; suboptimal from 800 to 1,000 and 1,400 to 2,000, lower than 500 and higher than 2,000, not suitable, because with such precipitation less irrigation is required (Figure 3).

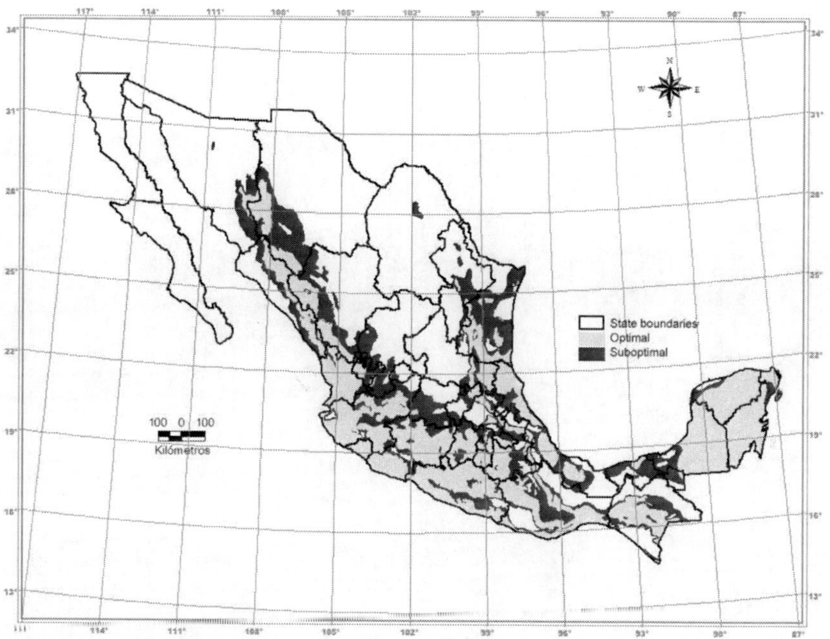

Figure 3. Optimal and Suboptimal Precipitations in Mexico.

Climate

The Stevia plant is subtropical from a jungle in high Paraná, native to northwestern province of Misiones, in Paraguay. The climates that are considered best are sub-humid temperate Aw and Cw and Cs. Such climates in Mexico are located in the region of the Yucatan Peninsula into the Gulf and the Pacific (Figure 4).

Temperature

Plants are not able to self-maintain a constant temperature, so temperature changes affect their growth and development, they are poikilothermic, but this does not mean that their temperature is the same as the environment, there can be differences.

It is certain that changes in temperature of the environment generate variations in the temperature of the plant.

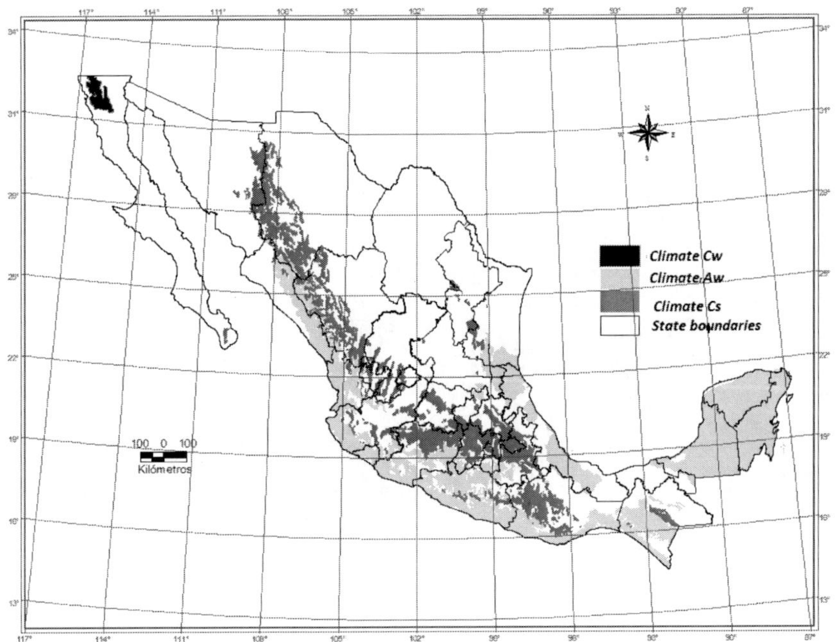

Figure 4. Optimal Climates for S. rebaudiana Bertori in Mexico.

The optimum temperature for growth of Stevia is 18 to 30° C with an average of 24° C; suboptimal of 15 to 18° C and 30 to 40° C. The extreme thermal limits are -6° C and 43° C (Figure 5).

Relative Humidity

The percentage of relative humidity for Stevia should be less than 85%. This factor directly influences the air temperature, the soil temperatures and the water vapor content in the atmosphere; it is also a determining factor in the incidence of diseases.

Brightness

Photoperiods (long day lengths) increase internode length, leaf area and dry weight that accelerate the appearance of leaves. Dry matter was reduced to half with short day photoperiods. The critical photoperiod for development of stevia is 13 hours, but there is considerable genetic variability among ecotypes. The most suitable soils for *S. rebaudiana* are the Franks, sandy loam and sandy clay loam, with regular rate of humus. It adapts accordingly to well-drained soils, but not in places with excessive moisture. It thrives effectively in acid soil pH, but grows well between 6.5 and 7.5, provided they are not saline. Considered as optimal for Mexico; Luvisols, Nitisols, Regosols and Fluvisols soils; suboptimal Leptosols (formerly Rendzinas and Lithosols) and Cambisols with good drainage; and as unfit Gleysols, Vertisols, and Lithosols Solonchaks.

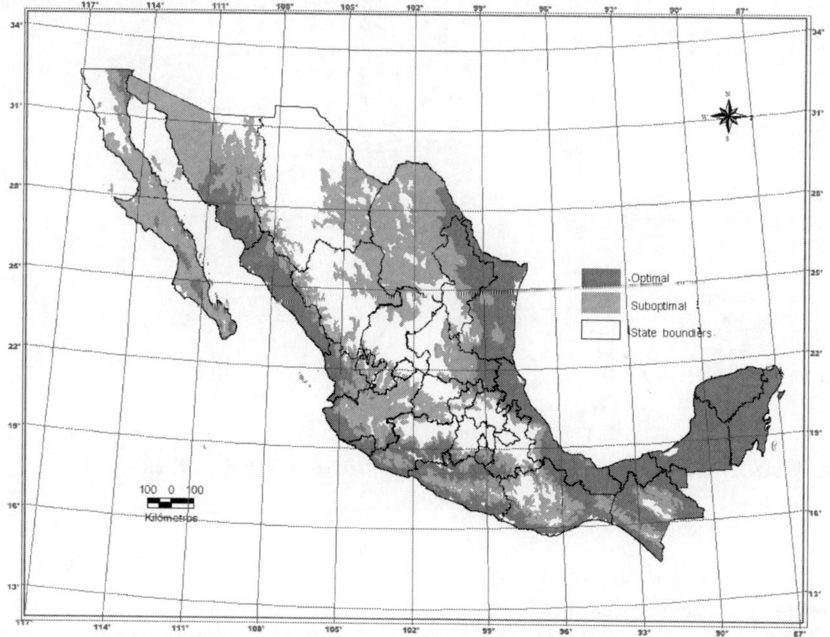

Figure 5. Optimal and Suboptimal Average Temperature in Mexico.

The low light condition in the tropics makes the Stevia plant have shorter cycles when blooming than at its origin, this cycle is between 45-60 days, depending on the conditions of precipitation, temperature and luminosity where it is located.

Soil

The best soils are mainly distributed in all the Gulf States and regions of the central Pacific and South Pacific (Figure 6).

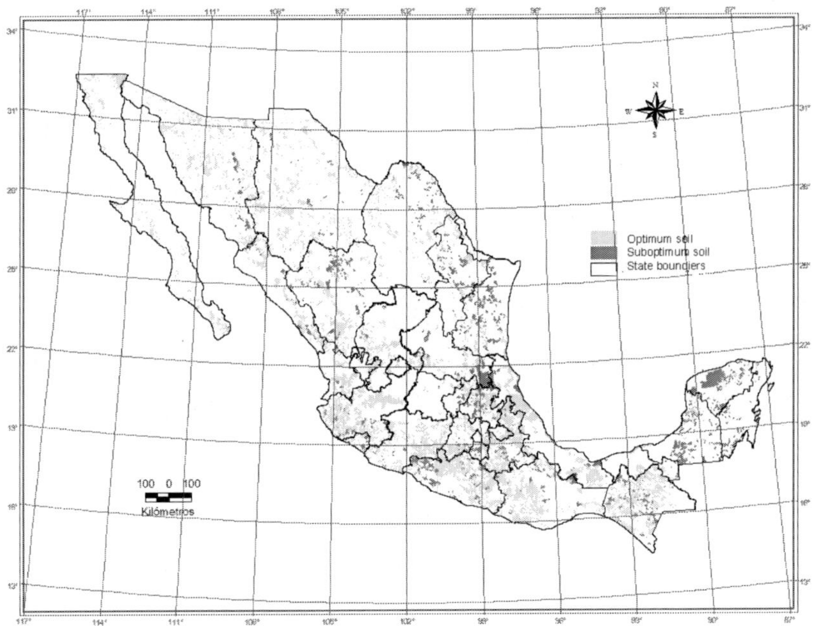

Figure 6. Space Distribution of optimal and suboptimal soil in Mexico.

The Productive Potential of *S. rebaudiana* under Irrigation in Mexico

The definition of productive potential is the spatial delimitation of geographical areas or agro ecological zones, where it is feasible, with a higher probability of success, the production of various agricultural, livestock and forestry species, with little or no damage to the environment.

The agro-ecological zones are areas where the soil is divided into smaller units with similar or homogeneous characteristics as much as possible, in relation to their suitability for the potential production of certain plant species. The details of these studies depend on the objective and therefore, the scale to be used. The present study is considered one of great vision in locating potential areas in Mexico for *S. rebaudiana* production.

Optimum (high) potential zones under irrigated conditions in Mexico are mainly distributed in the Pacific states such as Sinaloa, Nayarit, Jalisco, Colima, Michoacán, Guerrero, Oaxaca and Chiapas. In the Gulf of Mexico there are some optimal and suboptimal in the states of Tamaulipas, Veracruz and of less potential in Tabasco.

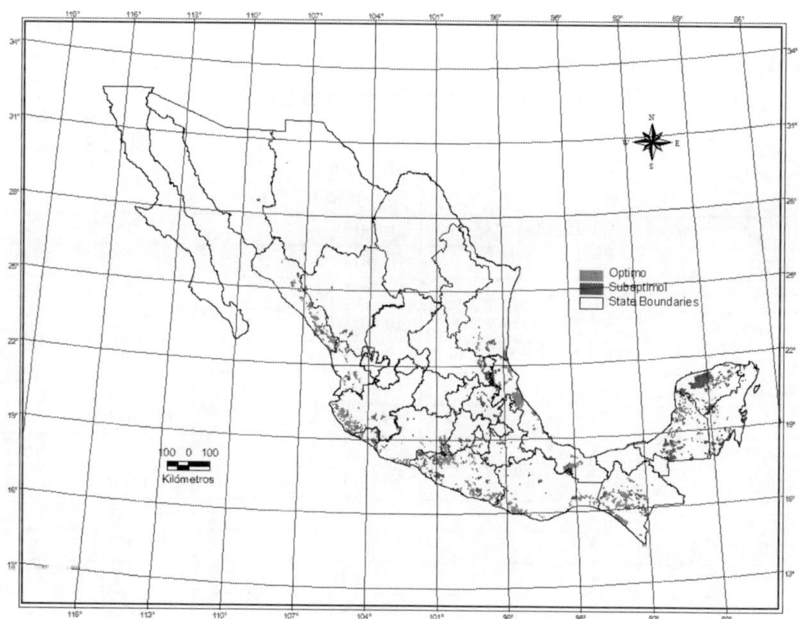

Figure 7. Distribution of potential areas for the Production of stevia in Mexico.

In the case of the Yucatan Peninsula, this area is mainly of medium-potential areas and some of optimum potential especially in the north-central state of Campeche and in the East and South of Yucatan (Figure 7 and Table 2). The areas with optimal potential are identified in figure 7 in Green, and suboptimal ones in red.

Sweetener Potential of *S. rebaudiana*

The quality of *S. rebaudiana* can be determined by the color of the leaves, and the contents of two of its main stevia glycosides: Stevioside and Rebaudioside-A. In general it can also be detected by the sweetness of the product.

It is cited, that out of the *S. rebaudiana* leaves that are produced in the world, the best quality leaves derive from South America and Mexico and contain 12 to 13% stevia glycosides. Leaves of lower quality are produced in China, because the content of stevia glycosides is only 5 to 6%. The content of steviol glycosides in the leaf depends on crop management, growth period and harvest (Shanks et al., 2004).

Table 2. Optimal and suboptimal surface (Ha) for *Stevia rebaudiana* Bertoni, by state distribution in Mexico

State	Optimal Conditions (Ha)	Suboptimal Conditions Ha)	Total (Ha)
Sinaloa	284,189	26,605	310794
Nayarit	252,574	13,302	265876
Jalisco	252,650	66,512	319162
Colima	157,830	39,908	197738
Michoacán	189,465	39,908	229373
Guerrero	410,320	106,420	516740
Oaxaca	252,659	39,908	292567
Chiapas	347,342	93,118	440460
Yucatán	94,729	266,050	360779
Campeche	157,975	159,630	317605
Quintana Roo	31,577	133,025	164602
Veracruz	221,036	53,210	274246
Puebla	0	39,908	39908
Tamaulipas	189,540	13,302	202842
Otros	315,765	239,445	555210
Total	3 157,651	1 330,251	4 487,902

Table 3 shows the content of Stevioside and Rebaudioside −A of the four materials studied. The analysis revealed statistically significant differences (P > 0.05) in the content of sweeteners among the materials studied.

Table 3 shows that the native material showed the highest concentration of stevioside, followed by Criole of Paraguay and Criole SM1, which were similar, and Morita II had the lowest concentration.

As for the content of Rebaudioside-A, it was found to be statistically equal in the varieties of SM1, Paraguay and Criollo varieties, and Morita II; this last one was the one with the highest content.

Concentrations of Stevioside and Rebaudioside-A found in the Criollo materials, Criollo Paraguay and SM1 were similar to those reported by other authors, as Kolb et al (2001), who reported content of 3.78 -. 9.75% Stevioside and 1.62 - 7.27% of Rebaudioside-A in S. rebaudiana leaves from Argentina, and Jackson et al (2009) with content of 4-14% Stevioside and 2-4% of Rebaudioside-A.

Table 3. Stevioside and Rebaudioside-A Content in the Varieties of Stevia

Variety	Stevioside g / 100 g de leaves Average ± Standard deviation	Rebaudioside-A g / 100 g de leaves Average ± Standard deviation
Criole	$7.15^c \pm 0.70$	$2.93^a \pm 0.27$
Criole of Paraguay	$5.09^b \pm 0.04$	$2.42^a \pm 0.08$
SM1	$5.03^b \pm 0.10$	$1.72^a \pm 0.09$
Morita II	$1.92^a \pm 0.43$	$10.12^b \pm 2.11$

[a, b, c] different literals in the same Column indicate statically significant difference.

Table 4. Steviosido and Rebaudioside-A in the Morita II variety by state

State	Stevioside g / 100 g leaves Average ± Standard deviation	Rebaudioside-A g / 100 g leaves Average ± Standard deviation
Chiapas	2.00 ± 0.02	11.00 ± 0.27
Quintana Roo	2.18 ± 0.40	9.14 ± 0.98
Veracruz	1.94 ± 0.07	7.79 ± 0.27
Yucatán	1.79 ± 0.48	10.80 ± 2.37

Although no differences were found in between states. The Morita II variety from Chiapas and Yucatan show a higher content of Rebaudioside-A.

It has been reported that Stevioside is the compound with the highest concentration among all the glycosides, as was observed in the materials studied, except for the Morita II.

The Morita II has shown different behavior as reported, and presented a lower content of Stevioside than Rebaudioside-A. This is an advantage because Rebaudioside-A is a less bitter glycoside and has a higher sweetening power (250-450) than Stevioside (150-300) (Jackson et al, 2009). Materials with these characteristics have a higher potential for the extraction industry as sweeteners. Because not all the varieties were produced in all four states being analyzed, the statistical analysis was only done considering the Morita II variety.

No statistically significant differences (p<0.05) were found between the mean concentrations of both Stevioside and Rebaudioside-A leaves of Morita II between states. In Table 4, we can see the concentrations found of glycoside between states.

Because a larger number of samples Morita II in Yucatan were obtained, a statistical analysis was performed to assess the behavior of this material in the different places where it was planted. The results indicated statistically significant differences (P> 0.05) between sites. The results are presented in Table 4.

Regarding the content of Stevioside, both Mocochá and Muna sites presented lower concentrations, while Tizimín and Baca sites showed the highest ones. There was no pattern or trend related to the type of soils that might have led to such results since Tizimín and Baca, with different and contrasting soils, showed the same concentrations of Stevioside as found for Rebaudioside-A.

Generally it can be seen that at the site of Muna the varieties obtained lower levels of Stevioside and higher levels of Rebaudioside-A, a desirable condition. However, by adding the two components a higher concentration was observed in the leaves of Morita II obtained in Baca, Yucatan (Table 5 and 6).

Regarding the commercial products (crystals) analyzed and identified as sweeteners 1 and 2, it was found that sweetener 1 had a concentration of Stevioside of 2.68% and 3.36% of Rebaudioside-A.

Table 5. Stevioside and Rebaudioside-A in the Morita II variety in four sites of Yucatán

Material	Stevioside g / 100 g leaves Average ± Standard deviation	Rebaudioside-A g / 100 g leaves Average ± Standard deviation
Baca	$2.52^c \pm 0.07$	$12.65^b \pm 0.14$
Mocochá	$1.42^a \pm 0.16$	$8.82^a \pm 2.30$
Muna	$1.52^a \pm 0.15$	$12.28^b \pm 0.78$
Tizimín	$2.23^b \pm 0.06$	$7.59^a \pm 0.03$

[a, b, d] Different literals in the same column represent significant different statistics.

Table 6. Stevioside and Rebaudioside-A content in the different varieties, states and sites of stevia

State	Site	Location	Variety	Stevioside g/100g hoja	Rebaudioside-A g/100g hoja	Total
Chiapas	Huehuetan	Rancho La Chinita	Morita II	2.01	11.01	13.01
Q. Roo	Bacalar	Stevia maya	Criole	7.15	2.93	10.08
Q. Roo	Chunhuhub	Unidad de Riego	Morita II	2.51	8.28	10.79
Q. Roo	Bacalar	Stevia maya	Morita II	1.85	10.00	11.85
Q. Roo	Chunhuhub	Unidad de Riego	Criole II*	5.14	2.51	7.66
Q. Roo	Bacalar	Stevia maya	SM_1	5.03	1.72	6.74
Veracruz	Cotaxtla	Campo Experimental	Morita II	1.94	7.79	9.73
Yucatán	Baca	Kacabchen	Morita II	2.52	12.65	15.18
Yucatán	Muna	CE Uxmal (1)	Morita II	1.50	11.60	13.10
Yucatán	Mocochá	CE Mocochá (1)	Morita II	1.28	10.81	12.09
Yucatán	Tizimín	Rancho La Montaña	Morita II	2.23	7.59	9.82
Yucatán	Mocochá	CE Mocochá (2)	Morita II	1.55	6.83	8.38
Yucatán	Muna	CE Uxmal (2)	Morita II	1.48	12.85	14.33
Yucatán	Muna	CE Uxmal (3)	Morita II	1.78	13.31	15.09
Yucatán	Muna	CE Uxmal (4)	Morita II	1.33	12.74	14.07

*Criole of Paraguay.

A commercial form can be a white powder with a sweet taste, but a very bitter aftertaste due to the concentration of Stevioside. Sweetener 2 was found in a presentation of sweet white crystals without the bitter taste, this due to the low concentration of Stevioside (0.3%) and also Rebaudioside-A (1.99%); this sweetener was labeled with a content of sucrose, consequently diminishing its properties as a natural non-caloric sweetener.

The Medical Potential of *S. rebaudiana*

For hundreds of years, the Guarani Indians of Brazil and Paraguay have used stevia leaves as a sweetener and in herbal medicine as a cardio tonic, for obesity, hypertension, and intestinal problems and for reducing the levels of uric acid.

Besides being a sweetener, stevia is considered (in Brazilian herbal medicine) as hypoglycemic, hypotensive, diuretic, cardio tonic, and tonic. The leaf is used for diabetes, obesity, dental cavities, hypertension, fatigue, depression, sweet cravings, and infections. The leaf is used in traditional medical systems in Paraguay for the same purposes as those in Brazil.

In Mexico, obesity is the major modifiable risk factor for the development of chronic non-transmissible diseases such as diabetes mellitus and cardiovascular disease (the two leading causes of death in Mexico), among other complications (World Health Organization, 2000 2003).

Stevia can play an important role in addressing some of the chronic diseases of the Mexican population, but to really make this an alternative focused studies are needed as they are doing countries such as Japan, Paraguay, Brazil, Denmark and Australia among others.

Information on the magnitude of the obesity problem in Mexico and two major non-transmissible chronic diseases such as diabetes and hypertension, as well as some studies that have been focused on this area are presented below.

Overweight and Obesity

The rate of obesity in Mexico is one of the highest in the world. During the last 12 years, from 2000 to 2012, the combined prevalence of the overweight and obese population increased 15.2%. Males increased 16.8% while females increased 13.9%.

By 2012, 26 million Mexican adults were overweight and 22 million were obese. These figures clearly indicate a major challenge for the health sector in terms of promoting healthy lifestyles in the population and development of public policies to reverse the obesogenic environment characterized by greater access to foods and beverages with high numbers of calories or high energetic levels, larger portions of food, sedentary lifestyles and an environment of constant promotion of the consumption of unhealthy products.

In 2008 in Mexico, 42,000 million Mexican Pesos were spent for obesity campaigns. This is equivalent to 13% of the total budget targeted to health costs (0.3% of GDP, Gross Domestic Product). If preventive measures are not applied directly to the costs to treat obesity, in the year 2017, it will be in the order of 101,000 million Mexican pesos; 101% more than the estimated costs in 2008. The indirect costs related to control the consequences of obesity:

hypertension, diabetes mellitus type 2, cardiovascular disease, breast and colorectal cancer could increase up to 292% between 2008 and 2017; from 25,000 to 73,000 million (Gutierrez et al., 2012).

For obesity and overweight problems, *S. rebaudiana* can play an important role because it is completely calorie-free. Furthermore, it is highly recommended for weight loss because it reduces anxiety for food and decreases the need to eat sweets.

Diabetes

Estimates from the World Health Organization (WHO) indicate that the number of people with diabetes worldwide has increased from 30 million in 1995 to 347 million at present and it is estimated that by 2030 it will be 366 million. Therefore, diabetes is considered a public health problem causing a great impact in terms of economic, social and in quality of life, thus becoming a national priority.

The prevalence of previously diagnosed diabetes in the National Health and Nutrition Survey 2012 (ENSANUT) was 9.2% (6.4 million), higher than that observed in the ENSANUT 2006 when it was 7.3% (3.7 million) and the 2000 NHS 4.6% (2.1 million). The prevalence was higher in older age groups; however, in the group aged 40 to 49 years a 50% increase was observed by from 2000 to 2006 and 2012. In all three surveys the prevalence of previously diagnosed diabetes was higher in urban than in rural areas (Jimenez-Corona et al., 2013).

The impact of diabetes on the national health system justifies the development of an action plan, where *S. rebaudiana* can play an important role, making it convenient to carry out research.

Studies by the Department of Endocrinology and Metabolism in Aarhus University Hospital in Denmark showed that Stevioside (active principle of S. rebaudiana) directly acts by stimulating the pancreatic beta cells thus generating a considerable secretion of insulin (Jeppesen et al., 2000). The results of these medical tests indicate that Stevia could have an anti-hypoglycemic potential role in people with type 2-diabetes (non-insulin dependent).

A report in 2012 considered *S. rebaudiana* glycosides to be a good choice for diabetics to use as sweetener as it provides sweetness with minimal calories. It does not induce a glycemic response after ingestion. Some studies regard the anti-hyperglycemic action to yield contradictory results, but conclude that its effects can be considered positive or neutral (Loria, 2012).

Hypertension

Hypertension (HT) is one of the main risk factors for developing cardiovascular disease, strokes and kidney failure, which are major causes of death in Mexico (Stevens, 2008).

In just six years, from 2000 to 2006, the prevalence of hypertension increased to 19.7%. It affects 1 in 3 Mexican adults (31.6% Barquera et al., 2010).

The results of the ENSANUT 2012 show that of an estimated 22.4 million Mexican adults, aged 20 years or older who have high blood pressure, only 11.2 million have been diagnosed by a doctor. In Mexicans who have been diagnosed and are being treated, figures show that 5.7 million show blood pressure levels that may be considered appropriate, controlling their condition (Ministry of Health, 2012).

S. rebaudiana is a mild hypotensive, as it lowers the blood pressure that is too high and regulates heartbeat. It has vasodilator, diuretic and cardio tonic effects.

Studies conducted in 2000 by the Division of Cardiovascular Medicine in Taipei hospital, in Taiwan, showed that in a group of patients treated with Stevioside, after three months a marked hypotensive effect was observed. In conclusion, these studies found that Stevioside is well tolerated and effective and can be used as an alternative for hypertensive patients.

Other Medicinal Properties

Other medicinal properties of S. rebaudiana mentioned are: Anti rheumatic, Antimicrobial (stevia extract eliminated E. coli, salmonella, staphylococci, bacilli, and did not affect beneficial bacteria, indicating a selective action) dental decay or cavities, (compatible with fluorine, stops the growth of platelets and prevents caries), combats anxiety action on antioxidant nervous system (is important to note that the highest content of antioxidants is not in the leaves but the stems), skin effects (revitalizing epithelial cells, helps in quick healing of wounds), Anti-inflammatory, Anti-cancer and can be used for the treatment of diarrhea (www.ecoagricultura.com).

CONCLUSION

Regarding the productive potential of *S. rebaudiana* in Mexico there are enough optimal and suboptimal conditions for it to be considered as an important factor to increase production.

These areas are open to farming and this would involve a process of productive restructuration, especially in areas of temperate and subtropical climate, where economic feasibility should be considered.

Concerning the sweetening potential of *S. rebaudiana* in Mexico, varieties evaluated as Morita II presented more than 12% content of Rebaudioside-A, with a sweeter flavor than Stevioside, which also has a sweet flavor but has a bitter taste.

In Mexico there are few studies on the use of *S. rebaudiana* in health; however it is a species that can provide some solutions to serious health problems such as overweight, obesity, diabetes and hypertension. Due to the previous information, more specific scientific research supporting the effects on health should be conducted.

CONSULTATIONS ON INTERNET

www.alimentacion-sana.org
www.ecoagricultor.com
www.fao.org
www.jbb-stevia.com

REFERENCES

Baradas, M. W. (1994). Crop requirements of tropical in: handbook of agricultural meteorology. J. F. Griffiths Editor. Oxford Univ. Press New York.

Barquera, S., Campos-Nonato, I., Hernandez-Barrera, L., Villalpando, S., Rodriguez-Gilabert, C., Durazo-Arvizu, R., Aguilar-Salinas, C. A. (2010). Hypertension in Mexican adults: results from the National Health and Nutrition Survey 2006. *Salud Publica de Mexico,* 52 suppl. 1:S63-S71.

Benacchio, S. S. (1982). Some 58 species of agroecological crop production potential in the American Tropics requirements. FONAIAP-National

Agricultural Research Center. Ministry of Agriculture. Maracay, Venezuela.

Doorenbos, J. y A. H. Kassam. (1979). Effects of water on crop yields. FAO Irrigation and Drainage Paper No. 33. Rome, Italy.

ESRI. (1996). Arc View GIS, The Geographic Information System for Everyone. Environmental Systems Research Institute, Inc. 380 New York Street, Redlands, CA 92373 USA.

FAO. (1993). Ecocrop, Ecological requirements of plant species database. Rome, Italy.

Fischer J. C. (2013). Future of Stevia. In memory Extensive work in the VII International Symposium on Stevia. CIRSE-CECODET. INIFAP. Mérida, Yucatán.

Garcia, E. (1988). Modifications to System Köppen Climate Classification, UNAM. Mexico, DF.

Gonzalez, M. (1984). Plant species of economic importance in Mexico. Ed Porrúa. Mexico, D. F.

Gutierrez-Delgado, C., Guajardo-Barron, V., Alvarez Del Rio. (2012). Cost of obesity: Market failures and public policies for prevention and control of obesity in Mexico. Chapter 11. In: Dommarco Rivera JA, Hernandez Avila M, Aguilar Salinas C, Vadillo Ortega F, Murayama Rendon C. Obesity in Mexico: recommendations for state policy. UNAM. Mexico, D. F.

INIFAP-SARH.- (a). (1993). Determination of Productive Potential of Plants by District Rural Development Campeche, SARH - INIFAP. Campeche, Camp.

INIFAP-SARH.- (b). (1993). Determination of Productive Potential Plant Species for the state of Yucatan, SARH - INIFAP. Merida, Yucatan.

INIFAP-SARH.- (c). (1993). Determination of Productive Potential of Plants by District Rural Development in Yucatan, SARH - INIFAP. Merida, Yucatan.

Jackson, A. U., Tata, A., Wu, Ch., Perry, R. H., Haas, G., West, L., and Cooks, G. (2009). Direct analysis of Stevia leaves for diterpene glycosides by absorption electrospray ionization mass spectrometry. *Analyst*, 134(5), 867-874.

Jeppesen, P. B., Gregersen, S., Poulsen, C. R., Hermansen, K. (2000). Stevioside acts directly on pancreatic cell to secret insulin: actions independent of cyclic adenosine monophosphate and adenosine triphosphate-sensitive K+-channel activity. *Metabolism,* 49 (2), 208-214.

Jimenez-Corona, A., Aguilar-Salinas, C. A, Rojas-Martínez, R., Hernandez-Avila, M. (2013). Diabetes mellitus type 2 and frequency of actions for prevention and control. *Salud Publica de Mexico*, 55, Suppl. 2, S137-S143.

Kolb, N., Herrera, J. L., Ferreyra, D. J., and Uliana, R. F. (2001). Analysis of Sweet Diterpene Glycosides from Stevia rebaudiana: Improved HPLC Method. *Journal of Agriculture and Food Chemistry*, 49 (10), 4538–4541.

Loria, K. V. (2012). Scientific Report. Stevia and its role in health. Scientific Report prepared in collaboration with Truvia. Buenos Aires, Argentina.

Medina, G. G., Ruíz, C. A. y Martinez, P. R. A. (1998). The climates of Mexico. Regional Center Pacific Research Center, INIFAP, SAGAR. Guadalajara, Mexico.

Ramirez, J. G. (1995). Potential areas for oil palm cultivation in Campeche. INIFAP - Campo Experimental Edzna.

Ramirez, J. G. (2000). Geographic Atlas of the State of Yucatan. Interactive CD-SISIERRA INIFAP-CONACYT-Fundación Produce Yucatán AC, Field Experimental Mocochá. Mocochá, Yucatán.

Ramirez, J. G., Gongora, G. S., Miranda, P. L. (2006). Strategic Study of the Agroindustrial Chain: Habanero Chili. Interactive CD. INIFAP - ENPRODAY. Field Experimental Mococha. Mococha, Yucatan.

Ramirez, J., G., Moguel, O. Y., Aviles, B. W., Gongora, G. S. (2012). Final Project Report: Validation, Development and Diffusion of Technology Stevia (Stevia rebaudiana Bertoni) under irrigated conditions in the Southeast of Mexico. Unpublished.

Ruiz, C. J. A. et al. (1999). Requirements Agroecologicos Crop. Technical Paper No. 3. Pacific Research Center Regional Center. INIFAP. SAGAR. Guadalajara, Jalisco. Mexico.

Shanks, T., Timcke, K., Krigbaum, J y Uno, J. (2004). Market Perspectives, Development and Potential Applications and International regulations in Paraguay Stevia. USAID-Paraguay Sell. Asuncion, Paraguay.

Secretaria de Salud (a). (2012). Hypertension in Mexican Adults: Importance of Timely Diagnosis Improve and Control, in: National Health and Nutrition Survey 2012. National Institute of Public Health of Mexico. Cuernavaca, Morelos. Mexico.

Secretaria de Salud (b). (2012). Obesity in Adults: The Challenges of Downhill in: National Health and Nutrition Survey 2012. National Institute of Public Health of Mexico. Cuernavaca, Morelos. Mexico.

Secretaria de Salud (c). (2012). National Results from the National Health and Nutrition Survey 2012. National Institute of Public Health of Mexico. Cuernavaca, Morelos. Mexico.

Stevens, G., Dias, R. H., Thomas, K. J. A., Rivera, J. A., Carvalho, N., et al. (2008). Characterizing the epidemiological transition in Mexico: National and subnational burden of diseases, injuries, and risk factors. *PLoS Med.*, 5 (6), e125.

Warrington, I. J. and E. T. Kanemasu. (1983). Corn growth response to temperature and photoperiod. I. Seedling emergence, tassel initiation and anthesis. *Agronomy Journal*, 75, 749-754.

Woelwer-Rieck U., Lankes Ch., Wawrzun A., Wu M. (2010). Improved HPLC method for the evaluation of the major steviol glycosides in leaves of Stevia rebaudiana. *European Food Research and Technology,* 231, 581–588.

World Health Organization/Food and Agriculture Organization. (2003). Diet, nutrition and the prevention of chronic diseases. Report of a joint WHO/FAO expert consultation. WHO Technical Report Series 916. Geneva, Italia.

World Health Organization. (2000). Obesity: Preventing and managing the global epidemic. Report of a WHO consultation. WHO (Technical Report Series No. 894), 203. Ginebra, Suiza.

In: Stevia rebaudiana
Editors: D. Betancur and M. Segura

ISBN: 978-1-63463-335-2
© 2015 Nova Science Publishers, Inc.

Chapter 2

DETERMINATION AND ISOLATION OF CHEMICALLY ACTIVE COMPOUNDS IN *STEVIA REBAUDIANA* LEAVES BY CONVENTIONAL SEPARATION INSTRUMENT TECHNIQUES

N. A. Samah[*] *and N. A. A. Zamri*
Faculty of Industrial Sciences & Technology,
Universiti Malaysia Pahang, Malaysia

ABSTRACT

There are many types of separation techniques used by scientists to determine and isolate the chemical components in *Stevia rebaudiana* leaves. Using approaches that range in complexity from simple to very complex, many researchers have contributed to the various methods used for the isolation and identification of active chemical compounds. The predominant compound in *Stevia rebaudiana* leaves is called steviol glycoside. There are nine active compounds in steviol glycosides: stevioside (SV), rebaudioside A (RbA), rebaudioside B (RbB), rebaudioside D (RbD), rebaudioside F (RbF), steviolbioside (StB), rubusoside (Rub), rebaudioside C (RbC), and dulcoside (DuA). Currently, the most conventional instrument used to analyze steviol

---

[*] Email: nurlin_abusamah@yahoo.com.

glycoside is the high-performance liquid chromatography (HPLC). Examples include the HPLC-mass spectrometer and HPLC-ultraviolet. These instruments are used to detect the entities and amount of active compounds in *Stevia rebaudiana* leaves. However, it needs an optimum procedure and critical analysis, especially of the parameter settings and types of columns used. Currently, there are no reported studies on the analysis of steviol glycosides through ultraviolet-visible spectroscopic methods. These methods are definitely simpler, cheaper, and eco-friendly due to the fewer chemicals and organic solvents used, and the mild pH control that is needed. The separation method to analyze the chemical composition and isolate the compounds will be discussed and compared. The steviol glycoside benefits humans and has remarkable potential as an intense high-potency sweetener together with its functional and health promoting properties.

INTRODUCTION

This chapter highlights the methods of separation and instruments used to perform separation and detection of nine active compounds, also known as steviol glycosides, in *Stevia rebaudiana* leaves. Steviol glycoside is obtained from the *Stevia rebaudiana* Bertoni leaves, and is a natural sweetener that can replace sucrose, glucose, and fructose obtained from honey, maple sugar, and table sugar. Sucrose, glucose, and fructose are natural sweeteners that give high-calorie value to the body system. Steviol glycoside represents nine molecules: stevioside, rebaudioside A, Rebaudioside B, rebaudioside C, rebaudioside D, rebaudioside F, dulcoside A, rubusoside, and steviolbioside (Figure 1). The 63rd Joint FAO/WHO Expert Committee on Food Additives (JECFA) reported that the commercially available extracts of *Stevia rebaudiana* leaves are stevioside and rebaudioside A. They are present in various amounts, ranging from about 10% to 70% for stevioside and 20–80% for rebaudioside A. According to the 73rd JECFA (2010), stevioside and rebaudioside A are the glycosides of principal interest due to their sweetening property. The rest are present in lower levels when compared to stevioside and rebaudioside A. Stevioside, or 13-[(2-O-β-D-glucopyranosyl-β-D-glucopyranosyl)oxy] kaur-16-en-18-oic acid, β-D-glucopyranosyl ester, is the main component that acts as the sweetener of *Stevia rebaudiana* leaves. Also, Rebaudioside A, or 13-[(2-O-β-D-glucopyranosyl-3-O-β-D-glucopyranosyl-β-D-glucopyranosyl) oxy] kaur-16-en-18-oic acid, β-D-glucopyranosyl (JECFA report 2007). Previous reports were performed to determine effective methods for the separation of the two active compounds from *Stevia rebaudiana* leaves extracts. Among the eight sweet diterpene glycosides, the most exhaustively studied was stevioside (Munish et al. 2012). The criteria for separation included that it must be efficient, simple, eco-friendly, and cheaper, as well as use a non-organic solvent during the analysis. There are many types of separation techniques that have been widely used, such as soxhlet extraction (Nurlin et al. 2013), ultrasonication (Nurul et al. 2014), solid phase extraction (Woelwer-Rieck et al. 2010), supercritical fluid extraction (Woelwer-Rieck et al. 2010), hydro distillation, and microwave extraction.

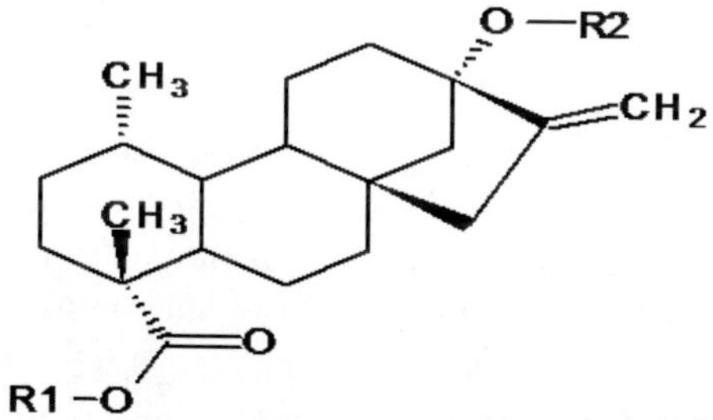

Name	R1	R2
Stevioside (SV)	Glucose	Glucose-1,2-Glucose
Rebaudioside A (RbA)	Glucose	Glucose-1,2-Glucose | 1,3-Glucose
Rebaudioside B (RbB)	H	Glucose-1,2-Glucose | 1,3-Glucose
Rebaudioside D (RbD)	Glucose-1,2-Glucose	Glucose-1,2-Glucose | 1,3-Glucose
Rebaudioside F (RbF)	Glucose	Glucose-1,2-Xylose | 1,3-Glucose
steviolbioside (StB)	H	Glucose-1,2-Glucose
Rebaudioside C (RbC)	Glucose	Glucose-1,2-Rhamnose | 1,3-Glucose
Dulcoside A (DuA)	Glucose	Glucose-1,2-Rhamnose

Source: Woelwer-Rieck et al. 2010.

Figure 1. Chemical structural of stevioside & rebaudioside A.

SEPARATION TECHNIQUES DEVELOPMENTS

In this sub-chapter the separation methods of steviol glycosides are discussed. There are many types of separation techniques used by researchers such as chromatography, spectroscopic and electrophoresis. Chromatography has been used for a century because of its functionality for the detection of analytes in samples at lower levels of detection. It also gives more accurate and precise data when compared to traditional methods. Chromatography has in common use of a stationary phase and a mobile phase. The components of a mixture are carried through the stationary phase by the flow of the mobile phase, and separation is based on the differences in migration rates among the mobile phase components (Skoog et al. 2004; Harvey et al. 2000). Most of the analyses conducted use a liquid as the mobile phase. Therefore, this technique is called liquid chromatography. Liquid chromatography is very significant in steviol glycosides separation because it has the ability to separate bulk compounds. However, the solid stationary phase allows for separation of the steviol glycoside compounds due to its migration rates in the stationary phase. Separation occurs in the column located in the oven. Several instrumental parameters need to be wisely adapted in order to isolate the active compounds from the matrix. These parameters include temperature of column, type of column, injection volume, mobile phase, mode of separation, and mode (isocratic gradient). In order to produce good separation, the peak that represents the analytes in the sample must be well-separated and cannot overlap with other peaks. The good separation is also called good resolution. The term resolution indicates whether the result is good or poor. From the observed chromatogram, the best analysis can be defined or vice versa. The chromatogram is a plotted graph of the detector's signal of elution time or volume (Harvey et al. 2000).

There are several important procedures that may provide good results when handled carefully and skillfully. Prior to running the high performance liquid chromatography (HPLC), the column needs to be flushed using a solvent or the mobile phase. The mobile phase needs to be prepared first and put into the solvent reservoir. Then, the mobile phase will go through the pump. In this compartment, the software, which is also called the solvent manager, controls the flow rate (volume/second unit) of the mobile phase. Next, the mobile phase passes through the auto-sampler (if present) or the injection valve, which then goes through the column. The column should be flushed using the mobile phase, which will reach the detector. Then, the mobile phase becomes waste. At the injection valve, the user injects a certain volume of the sample or standard solution using a syringe and the sample is carried out by the mobile phase into the column. Separation occurs in the column. Additionally, there is packing media content, also known as the stationary phase, in the column. For the most part, the stationary phase used depends on the type of column. The column is located in the oven, the temperature of which can be controlled. The mode of separation (isocratic or gradient) depends on the temperature. The isocratic mode is used if the same temperature is set throughout from start to finish. However, if the oven temperature will increase or decrease at certain retention times, then it is called temperature programming or the gradient mode. Subsequently, the analytes will be separated due to their migration rates in the stationary phase. The sample will then pass through the detector, which will give a signal to the processor. Towards the end, the peaks or chromatogram will be observed. Figure 2 and 3 shows the real HPLC in a lab.

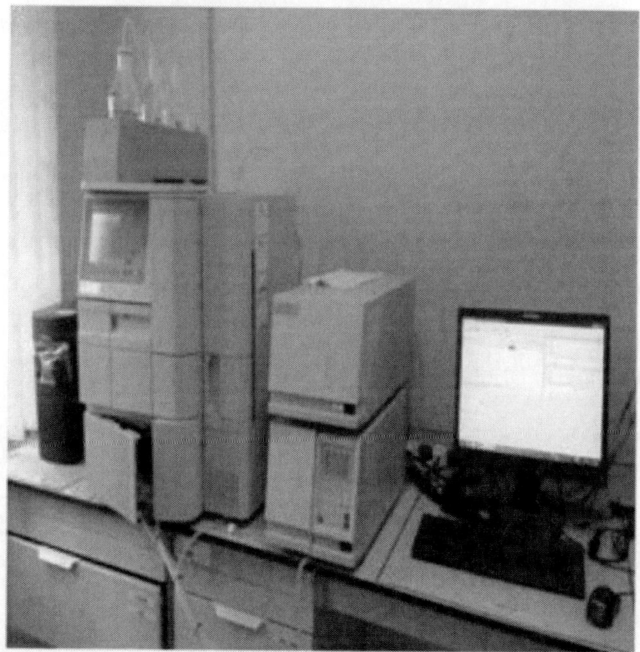

Source: Faculty of Industrial Sciences & Technology, Universiti Malaysia Pahang, Malaysia.

Figure 2. High Performance Liquid Chromatography.

A basic C_{18} column (Figure 3) was used. It is a good column due to its efficiency in separating the bulk compounds in a sample. Most instruments use this type of column to detect the analytes present in the sample. However, to develop a new method, other types of columns, such as an amino column, are preferred.

Source: Faculty of Industrial Sciences & Technology, Universiti Malaysia Pahang, Malaysia.

Figure 3. C_{18} column in HPLC Waters.

HPLC has been an important method for the separation of steviol glycoside (Rahul et al. 2013). It has been used as the main technique to develop improved methods. The principles of HPLC are based on chromatography techniques. It is highly sensitive and can detect trace levels of compounds parts at per billion (ppb). The hyphenated chromatography methods are widely used and are applicable to any analysis. These methods can be defined as separation techniques that include two or more separation techniques that consist of chromatography as well as the detector used. Many researchers have developed methods for isolating and purifying steviol glycosides. The table 1 shows the HPLC methods used for isolating active compounds.

Table 1. The hyphenated chromatography methods implemented in isolation and purification of steviol glycoside

Methods	Column used	Instrumental parameters	Findings
LC-MS-MS	C_{18} column (5 x 2.1 mm)	Ammonium acetate-ACN solvent system Isocratic elution Flow rate: 250 µL/minutes Column temperature:40°C Injection volume: 5 µL Reversed phase binary gradient	Active compounds profile observed at range 3.3-3.9 minutes for steviol glycoside. (Behnaz et al. 2012)
JASCO HPLC system	C_{18} column (150 x 4.6 mm)	ACN-water solvent system Rheodyne injector (20µL loop)	The peak gives significant at 1.958 minutes. Poor resolution observed. (Inamake et al. 2010)
Prep-HPLC	C_{18} column (150 x 4.6 mm)	ACN-water solvent system Column temperature: 25°C UV detector : $205 - 215$ nm Isocratic mode	The peak gives significant at 4.26 minutes with poor resolution. (Nurlin et al. 2013)
HPLC	NH_2 column	ACN:Na-phosphate buffer solvent system Column temperature: 27-28°C Isocratic mode UV detection at 210 nm Flow rate: 1.0 mL min^{-1}	Membranes filtration applied. The purity up to 97.66% with total yield steviosides was 9.05 g per 100 g stevia leaves. (Adari et al. 2012)

Table 1. (Continued)

Methods	Column used	Instrumental parameters	Findings
HPLC	C_{18} column (8 x 10 cm)	CH_3CN-water solvent system. Applied decolorization via electrolysis & demineralization by ion exchange. UV detection at 210 nm	8-10% yield of white to pale yellow color powder. 70-80% stevioside present in chromatogram. (John et al. 1987)
HPLC-UV	Luna HILIC column (250 x 4.6 mm)	ACN-water solvent system Isocratic condition Flow rate: 1mL/min Column temperature: 36°C Injection volume: 20μL	Good resolution between rebaudioside A and stevioside. HILIC column shows almost no bleeding. Rebaudioside A more stable against acid hydrolysis than stevioside. Stevioside strongly degraded. (Wolwer-Rieck et al. 2010)
HPLC-UV	Luna C_{18} column (250 x 4.6 mm)	ACN-water solvent system Isocratic condition Rheodyne injection valve Injection volume: 20μL Flow rate: 1mL/min UV detection at 210 nm	Retention time at 10 mins. The percentage stevioside in stevia powder and stevia leaf were 8.859% and 3.703% respectively (Annie et al. 2011)

Conventional instrument methods, such as HPLC-MS and GC-MS, are used in the pharmaceutical, healthcare, food processing, and nutrition industries. The focus active compounds in stevia extracts (rebaudioside A and stevioside), has been used for a long time in many food products; especially chocolates and drinks. This kind of sweetener gives zero calories, but is 300 times sweeter than sucrose (Nurlin et al. 2013, Adari et al. 2012, Mondaca et al. 2012). Additionally, it is a natural sweetener and does not cause any long term effects, such as failure of the kidneys, liver, and blood cells. Synthetic sweeteners, such as aspartame, are currently used commercially. The isolation and purification processes of steviol glycoside can be conducted using zero-organic solvents, but rather just deionized water as the media for separating it from the matrix. Adari et al. (2012) concluded that the consumption of organic solvents is not good for human health, and the process is not too expensive compared to other methods. They used membrane technology to produce high yields of stevioside, and it was considered safe since no organic solvents were used. Therefore, Adari et al. (2012) had developed a simple method that is inexpensive, as well as eco-friendly.

The table above shows the hyphenated chromatography methods for isolating the steviol glycoside compound from Stevia extracts. Most of the analyses were based on liquid chromatography using different types of columns or instrumental parameter settings. The amino column was suggested by the 73rd JECFA (2010) since it can produce the best resolution and results in stevioside that is more than 90% pure. The C18 column was applicable for most of the analyses. However, it produced poor resolution when compared to the amino column. The reversed mode of the column gives poor resolution and column bleeding might occur. The mobile phase used was included polar solvents and had the ability to attract the analytes from the matrix, thus allowing them to be detected after being separated in the column.

The principles of UV-Vis spectrophotometry are quite different from the chromatography techniques. Its level of detection is in the ppm range, and it is very user-friendly and does not produce any waste, which is similar to the chromatography techniques. Steviol glycoside compounds have no functional groups that absorb in the UV-Vis range, which is 200–700 nm. This is a challenge for researchers, who must develop methods prior to the analysis with the UV-Vis spectrophotometer (Figure 4). However, a few papers have reported on this issue and, therefore, it is explained extensively in this chapter. A UV-Vis spectrophotometer schematic diagram is shown in figure 5.

Source: Faculty of Industrial Sciences & Technology, Universiti Malaysia Pahang, Malaysia.

Figure 4. Ultraviolet-Visible Spectrophotometer, cuvettes compartment and sample cells

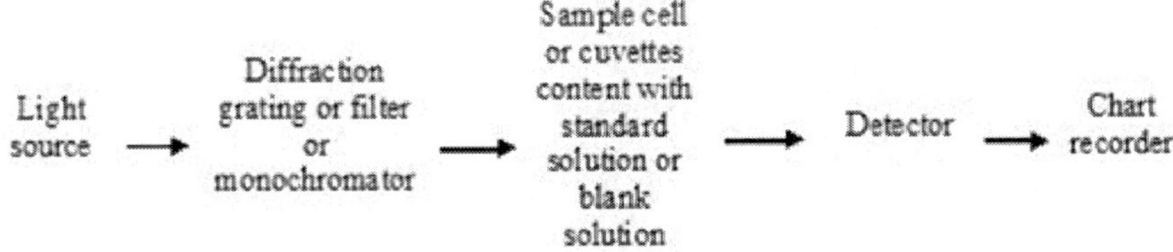

Figure 5. Schematic diagram of UV-Vis Spectrophotometer.

Spectroscopy or any measurement based on a light source that emits the electromagnetic radiation is absorbed by the analytes. The wavelength for ultraviolet light is between 200 and 400 nm, while the visible range is between 400 and 700 nm. The emitted light gives off energy at a certain wavelength, which is absorbed by the analytes or photons. Many had agreed to implement a formula that related energy (E) and wavelength (λ), and this formula guided the measurement in the detection of analytes' concentration (Equation 1). However, to determine the analytes' concentrations, this method uses the Beers' Law concept and results in one simple formula (Equation 2.2) (Skoog et al. 2004).

$$E = hv = \frac{hc}{\lambda} \qquad\qquad \text{(Equation 1)}$$

where * h is Planck constant value (6.626 X 10^{-34} J s) and c is the velocity of light (3.0 X 10^{8} ms^{-1}).

$$A = \varepsilon bc \qquad\qquad \text{(Equation 2)}$$

Where A is the absorbance value (ε) is the molar absorptivity, b is the path length of cuvettes, and c is the concentration of the analytes.

a) Pre-sonication assisted

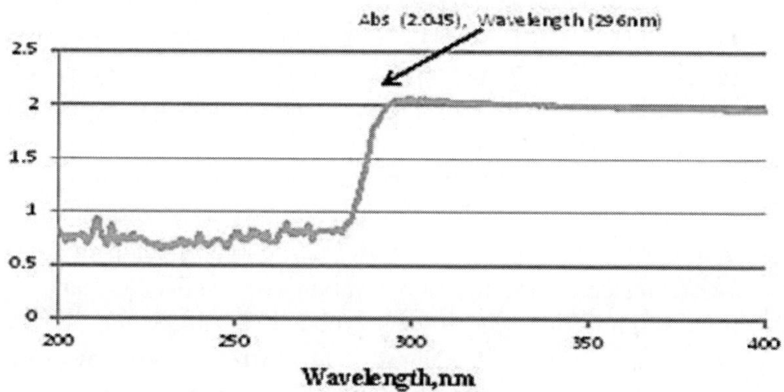

b) Post-sonication assisted

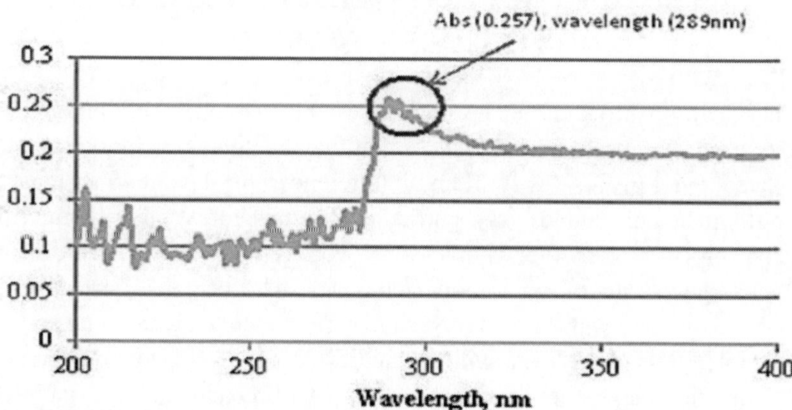

Source: Nurul, A. A. Z., 2014.

Figure 6. Signal of detector versus wavelength for (a) pre-sonication assisted and (b) post-sonication assisted in isolating stevioside in stevia extracts.

Purification of steviol glycosides was developed by Adari et al. (2012), stevioside produced with 97.66% purity (Adari et al., 2012). However, Nurul et al. (2014) reported a purity stevioside about 93% due to human error during the experimental process. This was due to difficulty maintaining continuous stirring, while chemicals are added during the period required for the crystallization process to occur. Also due to failures to wash the crystal well enough to eliminate any solution that may be attached to the crystals (Nurul et al. 2014).

UV detection at 210 nm or 200 nm after HPLC separation was often used for steviol glycoside determination. However, this is not very sensitive because the carboxylic acid and olefin moieties attached to the steviol backbone are weak UV absorbers (Wölwer-Rieck 2012). For analysis that does not involve any chromatography methods, the wavelength would be near 300 nm. This was agreed on by Inamake et al. (2010) and Nurul et al. (2014). Nurul et al. (2014) investigated the effect of pre- and post-sonication in stevia extracts for stevioside. While their observation was similar to Inamake et al. (2010), it was slightly different in peak production. At 290 nm, the slightly sharp peak was observed in the Nurul et al. (2014) study. However, both showed similar trends in the value of wavelength, which was approximately 290 nm qualitatively (Figure 6).

Clean-up Process in the Purification of Steviol Glycosides

The clean-up process can be defined as the method of eliminating the analytes from the interfering substances that may be contained in the samples. To isolate steviol glycosides from any interference, the stevia extract must be at a mild pH. This avoids any particular substance that may disturb the detection process after the sample has been running using instrumental methods or any kind of separation techniques. The chemical structures of steviol glycoside are well-known in such that the structures were quite similar, for example, rebaudioside A with rebaudioside B. This phenomenon also shows that the major stevioside compound and rebaudioside A gave the potential to be disturbed by other compounds that look similar. Therefore, this has made developing new methods that can simply differentiate the chemical structure for each compound more difficult. The best way to overcome this problem is using the Nuclear Magnetic Resonance (NMR). This technique uses high level energy until it can be absorbed on the atomic level. Additionally, it applies the spectroscopic methods and can produce results that include the position or arrangement of each atom in the compound so that the chemical structure of the steviol glycoside present in the sample can be observed. This method, however, is too expensive and requires a vast amount of knowledge about the atomic level, including the electrons and protons involved. Related to this topic, the novel structures of diterpene glycosides form stevia have been observed via ^{1}HNMR and ^{13}CNMR (Chaturvedula et al. 2011).

CONCLUSION

Currently, there are many separation techniques being developed. The most powerful tool in the isolation of steviol glycoside from stevia extracts is liquid chromatography due to its sensitivity and accuracy for the detection of trace compounds. However, there are also other methods (spectroscopic methods) that can be used to measure and detect the steviol glycoside compounds present in the samples. For this method, the analytes must be able to absorb light sources so that they can be detected. Therefore, prior to detection, the analytes must undergo preparation with additives or go through a clean-up process to isolate the steviol glycoside compound from the stevia extracts.

REFERENCES

[1]   Adari B. R., Goka, R. R., Prasad, E., Sudergopal, S. & Yerrapragada, V. Lakhsmi, R. (2012) Simple extraction and membrane purification process in isolation of steviosides with improved organoleptic activity. *Advances in Bioscience and Biotechnology, 3,* 327-335.

[2]   Adari, B. R., Goka, R. R., Prasad, E., Sudergopal, S. & Yerrapragada, V. Lakhsmi, R. (2012). An improvised process of isolation, purification of steviosides from stevia rebaudiana Bertoni leaves and its biological activity. *International Journal of Food Science & Technology, 47,* 2554-2560.

[3]   Annie, S., Vinit, P., Jay, B., & Saleemulla, K. (2011). Identification and estimation of stevioside in the commercial samples of stevia leaf and powder by HPTLC and HPLC. *International Journal of Pharmacy & Life Science, 2,* 1050-1058.

[4] Behnaz, S., Ramin, V., Randy, B., Ryan, W., & Daniel J. A. (2012). Large-scale profiling of diterpenoid glycosides from *Stevia rebaudiana* using ultrahigh performance liquid chromatography/tandem mass spectrometry. *Analytical and Bioanalytical Chemistry, 403,* 2683-2690.

[5] Chaturvedula, V. S. P., & Prakash, I. (2011) Structures of the novel diterpene glycosides from stevia rebaudiana. *Carbohydrate Research, 346,* 1057,-1060.

[6] Harvey, D. et al. (2000). Modern Analytical Chemistry (1st Edition) DePauw University, McGraw-Hill.

[7] Inamake, M. R., Shelar, P. D., Kulkarni, M. S., Katekar, S. M., & Tambe, R. (2010). Isolation and analytical characterization of stevioside from leaves of stevia rebaudiana bertoni (asteraceae). *International Journal of Research in Ayurveda & Pharmacy, 2,* 572-581.

[8] JECFA (Joint FAO/WHO Expert Committee on Food Additives) (2007) Evaluation of certain food additives and contaminants. Sixty-eight report of the Joint FAO/WHO Expert Committee on Food Aditives. World Health Organization (WHO), Geneva, 50-54.

[9] John, A., Duang, B., & Bela, T. (1987). Improved isolation and purification of stevioside. *Journal Science Social Thailand, 13,* 179-183.

[10] Mondaca, R. L., Antonio, V. G., Liliana, Z. B., Kong, A. Hen. (2012). Stevia rebaudiana bertoni, source of a high-potency natural sweetener: A comprehensive review on the biochemical, nutritional and functional aspects. *Food Chemistry, 132,* 1121-1132.

[11] Munish P., Deepika, S., Colin, J. B., & Tiwary, A. K. (2012). Optimisation of novel method for the extraction of steviosides from *Stevia rebaudiana* leave. *Food Chemistry, 132,* 1113-1120.

[12] Nurlin, A. S., Aliaa, D. A. Hisham, & Shaiful, A. R. (2013). Determination of stevioside and rebaudioside A in *Stevia rebaudiana* leaves via preparative High Performance Liquid Chromatography (prep-HPLC). *International Journal of Chemical and Environmental Engineering. 4,* No. 2.

[13] Nurul, A. A. Z. (2014). UMP technical writing. Determination of stevioside in *Stevia rebaudiana* leaves using sonication-assisted extraction via Ultraviolet-visible Spectrophotometer. Pahang, Malaysia.

[14] Rahul, S. P. et al. (2013). Sweeteners from plants-with emphasis on *Stevia rebaudiana* (Bertoni) and Siraitia grosvenorii (Swingle). *Analytical Bioanalytical Chemistry, 405,* 4397-4407.

[15] Skoog, D. A., Donald, M. W., James, H. F., & Stanley, R. C. (2004). Fundamentals of Analytical Chemistry (8th Edition). Belmont, Brooks/Cole Cengage Learning.

[16] Woelwer-Rieck U. (2010). Improved HPLC method for the evaluation of the major steviol glycosides in leaves of *Stevia rebaudiana*. *Europe Food Resource Technology, 231,* 581-588.

[17] Wölwer - Rieck U., Werner, T., & Andreas, W. (2010). Investigations on the stability of stevioside and rebaudioside A in soft drinks. *Journal of Agriculture Food Chemistry, 58,* 12216-12220.

[18] Wölwer-Rieck U. (2012). The leaves of stevia rebaudiana (Bertoni), their constituents and the analyses thereof: A review. *Journal of Agricultural and Food Chemistry, 60,* 886-895.

In: Stevia rebaudiana
Editors: D. Betancur and M. Segura

ISBN: 978-1-63463-335-2
© 2015 Nova Science Publishers, Inc.

Chapter 3

PHYTOCHEMISTRY OF *STEVIA REBAUDIANA*

Swati Madan[1], and Sumeet Gullaiya[1]*
[1]Amity Institute of Pharmacy, Amity University,
Noida, India

ABSTRACT

Stevia rebaudiana (Bert.) known as the sweet herb of Paraguay contains number of diterpene glycosides that taste sweet. The main constituents include stevioside and rebaudioside A. The plant is a natural low calorie sweetener which is 300 times sweeter than cane sugar. It is a natural and healthy alternative to sugar and artificial sweetener. The plant is well known for a wide range of application in the treatment of many diseases like diabetes, high blood pressure and weight loss in various Indian traditional system of medicine. The aim of the review is to outline the phytochemical constituents of *Stevia rebaudiana*, giving emphasis to the most significant components including steviosides and rebaudiosides.

INTRODUCTION

Stevia rebaudiana (Bert.) is a non-calorie natural sweetener belongs to the family Compositae (Asteraceae) native to Paraguay, whose leaves have been used for centuries as a sweetener (Viana and Metiver, 1980). *Stevia*

---

* E-mail: smadan3@amity.edu, (M):9891626956.

rebaudiana (S. rebaudiana) is a natural herb, low calorie sweetener. It is commonly known as Stevia, honey leaf. Stevia (Asteraceae) is a woody shrub that can reach 80 cm in height when it is fully matured (figure 1).

The Stevia genus comprises at least 110 species (Rajbhandari and Roberts, 1983) but there may be as many as 300. Its habitat extends from the southwestern United States to the Brazilian highlands (Soejarto et al., 1982). Different species of Stevia contain several potential sweetening compounds, with *Stevia rebaudiana (S. rebaudiana*) Bertoni being the sweetest of all (Soejarto et al., 1982; Kinghorn et al., 1984). The main sweet phytochemical component of the leaves of *S. rebaudiana* is stevioside (Genus, Safety of Stevia, 2007). Stevioside is the predominant sweet tasting glycoside about 5-22% of the dry leaves weight in *S. rebaudiana* and has been reported to be 250-300 times sweeter than sucrose (Soejarto et al., 1982). The yield of stevioside from dried leaves of *S. rebaudiana* can vary greatly, as it is depending upon the cultivar and growing conditions (Stevia The genus Stevia, 2002).

(Segura-Campos et al., 2014).

Figure 1. Stevia rebaudiana leaves.

The leaves of *S. rebaudiana* (Bertoni) accumulate at least eight steviol glycosides (SGs), the concentrations of which vary quite widely depending on the genotype and production environment. Besides, it also contains sterols, triterpenoids, flavonoids, caumarins (Samulsson and Gunnar, 1992; Kinghorn and Soejarto, 1985).

It has a wide range of applications in pharmaceutical and allied industries and also has been found to be calorie free and non toxic. Various disease conditions are associated with free radical oxidative stress (Devasgayam, 1996). Stevia leaf contain free radical scavengers and are well known for their therapeutic activity.

The plant is also reported to have contraceptive properties (Melis, 1999). It helps in the prevention of dental cavities (Fujita et al., 1970). The plant has been reported for antibacterial and antifungal properties (Cereda and Miranda, 2002; Jayaraman et al., 2008). Moreover, it has been reported that *S. rebaudiana* shows ability to maintain blood glucose level with glucose tolerance enhancement in diabetic patients (Curi et al., 1986). Therefore, it is an alternative as a natural sweetener to diabetics and others on carbohydrate-controlled diets (Gregersen et al., 2004). Stevia leaves are a good source of carbohydrate and other nutrients and hence a substitute for sugar in processed drinks (Gasmalla et al., 2014). Rebaudioside M could also be of great interest to the global food industry because it is well-suited for blending and is functional in a wide variety of food and beverage products (Prakash et al., 2014).

Stevia leaf extracts are used in Japan, Korea, and certain countries of South America to sweeten soft drinks, soy sauce, yogurt and other foods, whereas in United States they are used as dietary supplements (Soejarto et al., 1982).

CHEMICAL COMPOSITION

Stevia is one of the largest and most easily recognized genus in the tribe Eupatorieae (Robinson and King 1977). The complete chemical composition of Stevia species is not yet available.

However a variety of Stevia species has been tested for their chemical composition. In this chapter, the sweet and non-sweet chemical constituents of *S. rebaudiana*, as well as the compounds obtained by tissue culture techniques, are discussed.

The main phytochemical components of *S. rebaudiana* are:

1. Diterpeneoids

Ent-kaurene: Diterpenoids are the best known compounds isolated from *S. rebaudiana,* specifically the sweet tasting ent-kaurene glycosides (figure 2). Eight naturally occurring ent-kaurene glycosides have been described from *S. rebaudiana,* comprising stevioside, rebaudioside A-E, steviolbioside and dulcoside A (**1-6**), with stevioside (**1**) being the most abundant sweet-tasting compound in the leaves (Kinghorn et al., 1984). These eight sweet-tasting glycosides from *S. rebaudiana* contain a common aglycone called steviol (13-hydroxy-ent-kaur-16-en-19 oic acid) (**7**) and differ only in the glycoside.

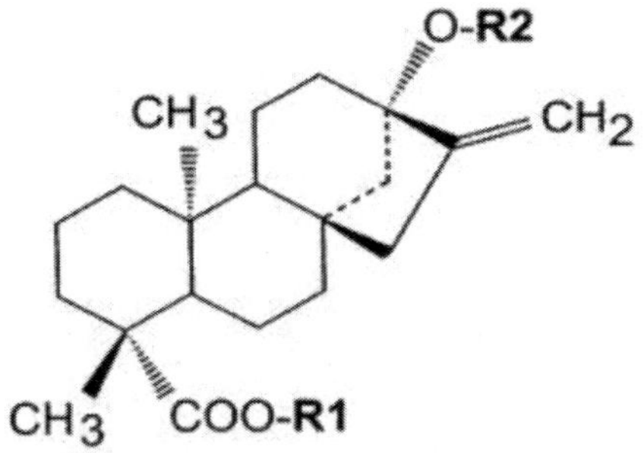

Cpd. No.	Compounds	R_1	R_2
1.	Stevioside	Glc	Glc-Glc (2→1)
2.	Rebaudioside A	Glc	Glc-Glc (2→1) | Glc (3→1)
3.	Rebaudiaoside C (= Dulcoside B)	Glc	Glc-Rha (2→1) | Glc (3→1)
4.	Rebaudioside D	Glc-Glc (2→1)	Glc-Glc (2→1) | Glc (3→1)
5.	Rebaudioside E	Glc-Glc (2→1)	Glc-Glc (2→1)
6.	Dulcoside A	Glc	Glc-Rha (2→1)

Glc= β-D-glucopyranosyl; rha= α-L-rhamnopyranosyl.

Figure 2. Structures of sweet-tasting *ent*-kaurenediterpenes isolated from *S.rebaudiana.*

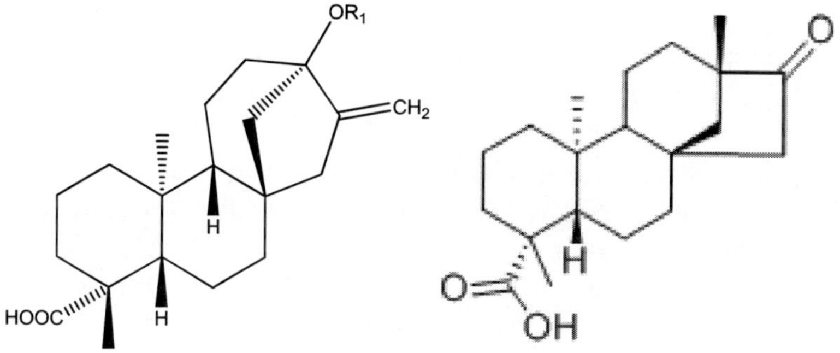

R1

7 Steviol

9 Steviolbioside

10 Rebaudioside B

8 Isosteviol

H

Glc-Glc (2→1)

Glc-Glc (2→1)

|

Glc (3→1)

Figure 3. Structures of derivatives of the S. *rebaudiana sweet ent* - kaurene diterpenoid constituents.

Stevioside is a glycoside with a glucosyl and sophorosyl residue attached to the aglyconesteviol, the end of the structure has a cyclopenatnoperhydrophenanthrene skeleton. The C_4 and C_{13} of steviol are connected to the β-glucosyl and β-sophorosyl group, respectively. Stevioside is obtained in the form of a white crystalline compound and is 100 to 300 times sweeter than table sugar (Stevia: Prospects, WHO, 2007; Goyal, 2010). A novel Enzymatic extraction method was also developed for the isolation of stevioside from S. *rebaudiana* leaves with cellulase, pectinase and hemicellulase, using various parameters, such as concentration of enzyme, incubation time and temperature (Puri et al., 2012). In another study, a new improvised process of extraction of steviosides from the stevia leaves in which the dry treated leaves were grounded, defatted, and extracted through pressurized hot water extractor (PHWE), followed by purification and concentration of the sweet glycosides through ultra (UF) and nano (NF) membrane filtration in obtaining high (98.2%) purity steviosides were studied (Rao et al., 2012). The yield of stevioside from dried leaves of S. *rebaudiana* can vary greatly, from about 5-22% of the weight of dry leaves, depending upon the cultivar and growing conditions (Kim and Dubois, 1991). Stevioside has also been found in the flowers of S. *rebaudiana* at lower concentrations (0.9% w/w) (Darise et al., 1983). Saponification of stevioside (**1**) with strong base yields steviolbioside (**9**). Tanaka and Coworkers reported the isolation of the novel-ent-kaurene *Rebaudioside A (2)* and *Rebaudioside B (10)* (Khoda et al., 1976). Rebaudioside A was obtained from a methanolic extraction of the leaves of *S. rebaudiana*, with a yield of 1.4% (w/w). The structure of rebaudioside A is the same as that of stevioside except that the sophorosyl residue is replaced by a glucosyl-(1-3)-sophorosyl residue. Rebaudioside A is the sweetest and the most stable glycoside. It is less bitter than stevioside. Rebaudioside E is as sweet as stevioside and rebaudioside D is as sweet as rebaudioside A, while the other glycosides are less sweet than stevioside. Hydrolysis of rebaudioside A with acid yields isosteviol (**8**) while hydrolysis of rebaudisoside A with hesperidinase yielded a partial hydrolysis product identified as rebaudioside B (**10**) (Khoda et al., 1976). Rebaudioside A is the sweetest of the ent-kaurene glycosides, approximately 350-450 times sweeter than sucrose, while rebaudioside B (**10**) is approximately 300-350 times sweeter than sucrose (Crammer and Ikan, 1987). Rebaudioside A is more pleasant tasting and more water soluble than stevioside and therefore it is better suited for use in food and beverages (Crammer and Ikan, 1987). Rebaudioside A has been identified in the flowers of *S. rebaudiana* at low concentration, 0.15% (w/w) (Darise et al., 1983). Stevioside A_3 is a synonym for rebaudioside A and stevioside A_4 is a synonym for rebaudioside C (Buckingham, 1987). Three minor ent-kaurene glycosides, rebaudiosides C-E (**3-5**) were isolated by Tanaka and co-workers (1982) from methanolic extract of the leaves of *S. rebaudiana*. (Sakamoto et al.,1977 a, b) having yields of 0.4%,0.03% and 0.03% w/w, respectively. The methanolic extract was recrystallized to remove stevioside (**1**), and the remaining extract subjected to repeated chromatography over silica gel to yield rebaudioside C-E (**3-5**) (Sakamoto et al., 1977a).The new compound was identified as 13-[(2-*O*-β-D-glucopyranosyl-3-*O*-β-D-glucopyranosyl-β-D-glucopyranosyl)oxy]*ent*-kaur-16-en-19-oicacid-(2-*O*-L-rhamnopyranosyl-β-D-glucopyranosyl) ester (**1**) on the basis of extensive spectroscopic (NMR and MS) and chemical studies from the leaves of *S. rebaudiana*. Two additional novel minor diterpene glycosides were also isolated from the commercial extract of the leaves of *Stevia rebaudiana*. (Prakash and Chaturvedula, 2013).

Dulcosides A and B were isolated from *S. rebaudiana* in minor quantities and they have been reported to taste moderately sweet (Kobayashi et al., 1977). The presence of these sweet-tasting ent-kaurene glycosides has been reported in the leaves, stems, and flowers of *S. rebaudiana*, but not in the roots (Tanaka, 1982).

S. rebaudiana is a short-day perennial composite, which when grow in long-day conditions, has increased levels of stevioside (**1**) (Metivier and Viana 1979). Post-harvest levels of stevioside and rebaudioside A (**2**) was examined by Cheng et al., (1981) and it was found that if *S. rebaudiana* is air dried for four to seven days to a moisture content of 15-20%, stevioside and rebaudioside content decreases significantly. The optimum temperature for the production of stevioside in greenhouse-cultivated *S. rebaudiana* plants has been reported to be 25°/20°C (day/night) (Mizukami et al., 1983). Seed propagation of *S. rebaudiana* plants result in a great variation in the levels of steviol glycosides (Tamura et al., 1984).With *in vitro* callus cultures of Stevia the production of measurable levels of the sweet-tasting glycosides did not occur (Handro et al., 1977, Wada et al., 1981).Within several years, however, investigators reported the accumulation of stevioside and rebaudioside A in callus culture of *S. rebaudiana* (Lee et al., 1982, Hsing et al., 1983). Other groups have reported the production of steviol glycosides in organ culture systems, including shoot-tip cultures (Yamazaki et al., 1991; Swanson et al., 1992).

2. Labdane

In addition to ent-kauren editerpenes, a number of labdane-type diterpenes have been isolated from *S. rebaudiana* and IR spectral analysis were also performed (Chaturvedula et al., 2012) (Figure 4).Two known labdane diterpenes, *jhanol (11)* and *austroinulin (12)* and one new diterpene, *6-O-acetylaustroinulin* (**13**), were isolated from a methanolic extract of *S. rebaudiana* leaves (Sholichin et al., 1980). The diterpenoid constituents have also been isolated from flowers that were jhanol (**11**), austroinulin (**14**) and 6-O-acetylaustroinulin (**13**), along with new labdane, *7-O- acetylaustroinulin* (**14**) with the yields of 0.04, 0.21, 0.18, and 0.08% w/w, respectively (Darise et al., 1983). A series of eight novel labdane-type diterpeneoids, *sterebins A-H* (**15-22**) have been identified from the leaves of *S. rebaudiana* (Oshima et al., 1986; 1988). Sterebin B was saponified with 1N potassium hydroxide to yield sterebin A (**15**). Sterebins A-D (**15-18**) have also been isolated from the leaves of *S. rebaudiana* in low yield of 0.001%, 0.0009%, 0.0003%, and 0.0004% w/w, respectively. *Sterebins A-D* are bisnorditerpenoids having C_{18} labdane skeleton and lacking C-14 and C-15 typical labdane skeleton. Bisnorditerpenoids are unusual compounds and sterebins A-D were the first to be reported with such highly oxidized B-rings (Oshima et al., 1986). The four additional novel labdane-type diterpenoids from *S. rebaudiana, sterebins E-H* **(19-22),** are not bisnorditerpenoids (Oshima et al., 1988).

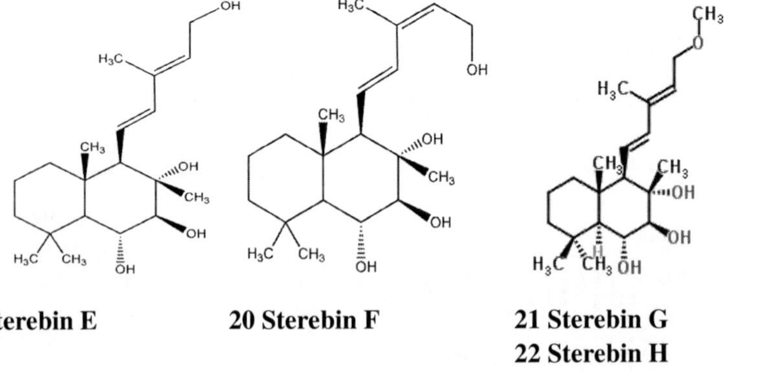

Cpd. No.	Compounds	R1	R2
12	Austroinulin	OH	OH
13	6-*O*-Acetyl-austroinulin	OAc	OH
14	7-*O*-Acetyl-austroinulin	OH	OAc

Cpd. No.	Compounds	R1	R2
15	Sterebin A	OH	OH
16	Sterebin B	OAc	OH
17	Sterebin C	OH	OAc
18	Sterebin D	H	OH

19 Sterebin E   **20 Sterebin F**   **21 Sterebin G**
                                                     22 Sterebin H

(*Sterebins G and H are C-14 epimers)

Figure 4. Structures of labdane-type diterpenes isolated from *S. rebaudiana*.

They have been reported to be isolated in low yield for sterebins E-H, 0.002%, 0.003%, 0.0002% and 0.0002% (w/w) respectively (Oshima et al., 1988).

3. Triterpenoids and Steroids

The known triterpenoid β-amyrin acetate (**23**) and three unidentified esters of lupeol were obtained from the ether-soluble portion of a methanolic extract of *S. rebaudiana* leaves (Sholichin et al., 1980). From methanolic extract of leaves of *S. rebaudiana*, lupeol ester, lupeol 3-palmitate (**24**) were isolated by GC-MS. β-sitosterol (**25**) and stigmasterol (**26**) comprising yield of 39.4% and 45.8% (w/w) of the total fraction was isolated from *S. rebaudiana* (D'Agostino et al., 1984).

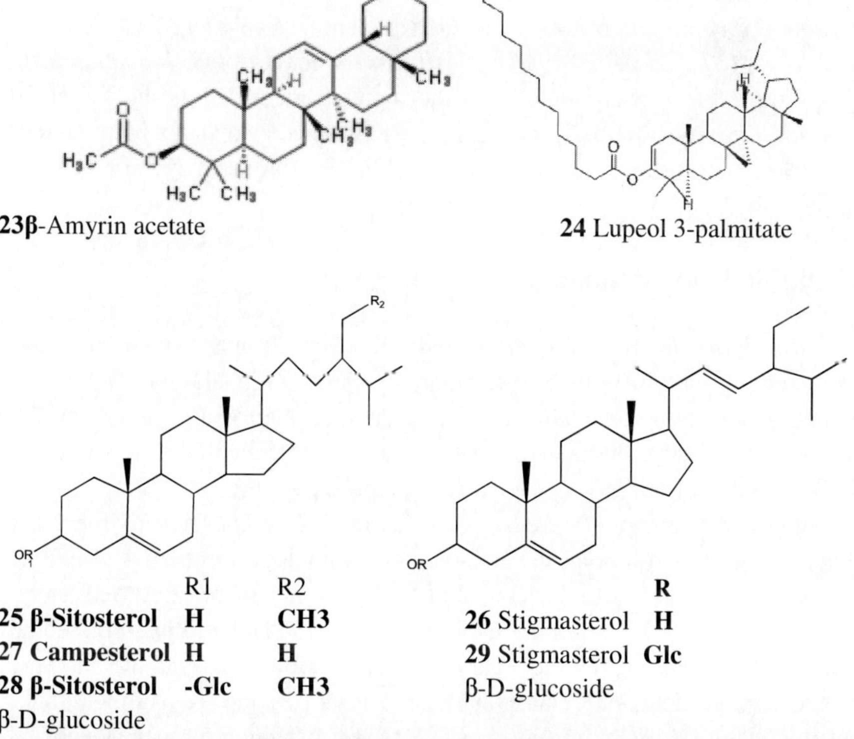

23β-Amyrin acetate

24 Lupeol 3-palmitate

	R1	R2		R
25 β-Sitosterol	**H**	**CH3**	**26** Stigmasterol	**H**
27 Campesterol	**H**	**H**	**29** Stigmasterol	**Glc**
28 β-Sitosterol	**-Glc**	**CH3**	β-D-glucoside	
β-D-glucoside				

Figure 5. Structures of triterpenes and sterols isolated from *S. rebaudiana*.

Campesterol **(27)** having a yield of 13.1% of the sterol fraction of the plant extract (D'Agostino et al., 1984) was also identified. The callus culture from the leaves of *S. rebaudiana* has also been shown to produce stigmasterol **(26)** (Nabeta et al., 1976).

Non-sweet steroid glycosides of *S. rebaudiana* were isolated by Matsuo et al., (1986) that is β-sitosterol-β-D-glucoside **(28)** and stigmasterol-β-D-glucoside **(29).**

4. Flavonoids

Six flavonoid glycosides were isolated from the ethyl acetate fraction of *S.rebaudiana* leaves, comprising apigenin 4'-O-glucoside **(30),** kaempferol 3-O-rhamnoside **(31),** luteolin 7-O-glucoside **(32)**, quercetin 3-O-arabinoside **(33)**, quercetin 3-O-rhamnoside (quercitrin) **(35)**. Rajbhandari and Roberts (1983) reported the isolation of methoxylated flavonoid, 5,7,3' trihydroxy-3,6,4' trimethoxyflavone (centaureidin) **(36)** from the chloroform extract of leaves of *S.rebaudiana*. Matsuo *et al*, (1986) identified apigenin 7-O-glucoside (cosmosiin) **(37)**, non-sweet glycosides from the leaves of *S. rebaudiana*. In cell culture of *S. rebaudiana* one of the pigment identified as the flavonoid glycoside *rutin* (quercetin 3-β-rutinoside) **(38)** has been isolated with a yield of 0.0007% of the fresh weight of the culture cells (Suzuki et al., 1976).

5. Volatile Oil Components

The volatile oils from the leaves and inflorescence of *S. rebaudiana* have been studied and a number of monoterpenes, sesquiterpenes, alkaloids, aldehydes and aromatic alcohols have been identified (Fujita et al., 1977; Martelli et al., 1985). Fujita et al., (1977) were able to identify 32 components of the essential oil of *S. rebaudiana*, including the sesquiterpenes β-caryophyllene, trans-β- farnesene, α-humulene, δ-cadinene, caryophyllene oxide, nerolidol and among the monoterpenes, linalool, terpinen-4-ol and α-terpineol were found as major constituents. In another study, the dried leaves of S. reabaudiana were extracted by steam distillation and two major essential oil components in the extract, caryophyllene oxide and spathulenol were isolated by column chromatography. Analysis of the remaining minor constituents by GC-MS showed over 100 peaks, of which 54 were identified (Martelli et al., 1985).

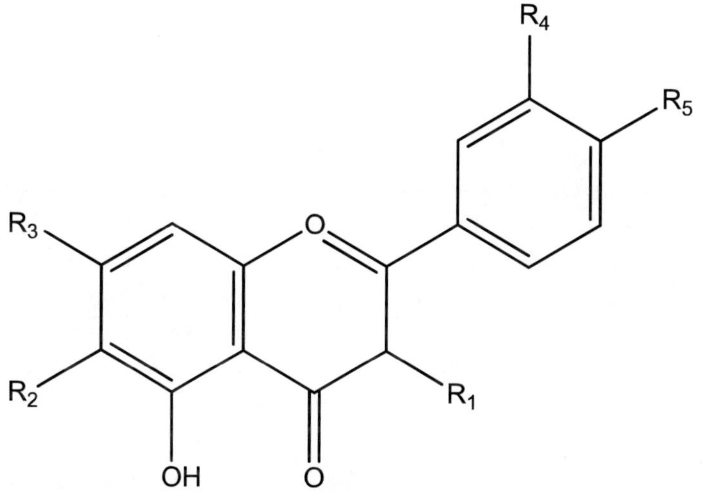

Cpd. No.	Compound	R_1	R_2	R_3	R_4	R_5
30	Apigenin4'-*O*-glucoside	H	H	OH	H	Glc
31	Kaempferol 3-*O*-rhamnoside	Rha	H	OH	H	OH
32	Luteolin 7-O-glucoside	H	H	Glc	OH	OH
33	Quercetin 3-O-arabinoside	Ara	H	OH	OH	OH
34	Quercetin 3-O-glucoside	Glc	H	OH	OH	OH
35	Quercetin 3-O-rhamnoside	Rha	H	OH	OH	OH
36	Centaureidin	OMe	OMe	OH	OH	OMe
37	Apigenin 7-O-glucoside	H	H	Glc	H	OH
38	Quercetin 3-O-rutinoside	Rut	H	OH	OH	OH

Glc=O-β-D-glucopyranosyl;   rha=O-α-L-rhamnopyranosyl;   ara=*O*-α-L-arabinopyranosyl.

Rut= 6-O- α-L-rhamnopyranosyl-D-glucopyranosyl.

Figure 6. Structures of flavonoids isolated from *S. rebaudiana*.

6. Other Constituents

A number of common phytochemical constituents have been identified in *S. rebaudiana*, such as the pigments of chlorophylls A and B and β-carotene (Cheng and Chang 1983). Various gums comprising 7-15% of the total extracted solids from *S. rebaudiana* solids were also identified. Tartaric acid was the major organic acid in an extract of *S. rebaudiana* with citric, formic, lactic acid in an extract of *S. rebaudiana* with citric, formic, lactic, malic and succinic acids also being identified (Cheng and Chang 1983). The phytohormone indole-3-acetonitrile has been reported from the seeds of *S. rebaudiana* and identified by retention time and color tests (Randi and Felippe, 1981). Unspecified tannins have also been reported in *S.rebaudiana* (Chung and Lee 1978). Inorganic compounds comprise approximately 13% of the total extractable portion from *S. rebaudiana*, with potassium being the major inorganic constituent of the Stevia extract (Cheng and Chang, 1983). Other inorganic substances identified from *S. rebaudiana* include calcium, iron, magnesium, phosphorus, sodium and zinc (Cheng and Chang, 1983).

CONCLUSION

The phytochemistry of *S. rebaudiana* has been studied intensely for decades now and many sweet and non-sweet compounds have been identified. Six new, naturally occurring sweet-tasting steviol glycosides have been identified from *S. rebaudiana*, with stevioside (**1**) being the most abundant, and rebaudioside A (**2**) being the second most abundant. The remaining four natural product ent-kaurene glycosides, rebaudiosides C-E (**3-5**) and dulcosideA (**6**), are present in lesser amounts. Due to their economic importance, attempts have been made to increase the yield of the steviol glycosides *in vivo* and also to produce these compounds *in-vitro* using various culture systems. In addition to the ent-kaurene glycosides, *S. rebaudiana* leaves contain a series of novel non-sweet labdane diterpenes, sterebins A-H (**15-22**). *S. rebaudiana* also elaborates a complex mixture of known sterols, triterpenoids, essential oils, flavonoids and other components.

REFERENCES

Buckingham, J. (Ed). (1997). Dictionary of Natural Products, Chapman 7 Hall, New York, CD-ROM, release 6:1.

Cereda-Garcia R., Miranda, P. (2002). HPLC isolation and structural elucidation of diastereomericniloylestertetrasachharides from Mexican scammony root. *Tetrahedron, 58*, 3145-3154.

Chaturvedula, V. S. P., Mubarak, C., Prakash, I. (2012). IR Spectral Analysis of Diterpene Glycosides isolated from *Stevia rebaudiana*. *Food and Nutrition Sciences*, 3, 1467-1471.

Cheng, T. E., and Chang W. H. (1983). Studies on the non stevioside components of Stevia extracts. National Science Council Monthly. *Taipei*, 11, 96-108.

Cheng, T. E., Chang, W. H. and Chang, T. R. (1981). A study on the post harvest changes in steviosides contents of Stevia leaves and stems. National Science Council Monthly. *Taipei*, 9, 775-782.

Chung, M. H. and Lee, M. Y. (1978). Studies on the development of Hydrangea and Stevia as natural sweetening products. *SaengyakHakhoechi*, 9, 149-156.

Crammer, B. and Ikan, R. (1987). Progress in the chemistry and properties of the rebaudiosides. In Developments in Sweetners-3, T.H. Grenby (Ed), Elsevier Applied Science, London, pp. 45-64.

Curi, R., Alvarez, M., Bazotte, R. B., Botion, L. M., Godoy, J. L., Bracht, A. (1986). Effect of *Stevia rebaudiana* on glucose tolerance in normal adult humans. *Brazilian Journal of Medical and Biological Research*, 19, 771-774.

D'Agostino, M., De Simone, F., Pizza, C. and Aquino, R. (1984). Sterols from *Stevia rebaudiana* Berroni. Bollettino-Societa Italiana di Biological Sperimantale 60, 2237-2240 [Chemical abstracts (1985) 102, 109851d]

Darise, M., Kohda, H., Mizutani, K., Kasai, R. and Tanaka, O. (1983). Chemical constituents of flowers of *Stevia rebaudiana* Bertoni. *Agricultural Biological Chemistry*, 47, 133-135.

Devasgayam, T. P. A., Kamat, J. P., Mohan, H., Kesvan, P. C. (1996). *Biochimica et Biophysica Acta*, 1282, 63-70.

Fujita, Hoehnea, Edahira. (1970). Safety utilization of Stevia sweetener. *The food Industry*, 82, 65-72.

Fujita, S. I., Taka, K. and Fujita, Y. (1977). Miscellaneous contributions to the essential oils of plants from various territories. XLI. On the components of the essential oil of *Stevia rebaudiana* Bertoni. *Yakugaku Zasshi*, 97, 692-694.

Gasmalla, M. A. A., Yang, R., Amadou, I., Hua, X. (2014). Nutritional Composition of *Stevia rebaudiana* Bertoni Leaf: Effect of Drying Method. *Tropical Journal of Pharmaceutical Research*, 13(1), 61-65.

Geuns, J. M. C. (2007). Safety of stevia and steviosides. *Recent Research Developments Phytochemistry*, 4, 75-88.

Gregersen, S., Jeppesen, P. B., Holst, J. J., Hermansen, K. (2004). Antihyperglycemic effects of stevioside in type 2 diabetic subjects. *Metabolism, Clinical and Experimental,* 53,73-76.

Goyal, S. K., Samsher, Goyal, R. K. (2010). Stevia (*Stevia rebaudiana*) a biosweetener: a review. *International Journal of Food Science and Nutrition,* 61(1), 1-10.

Handro, W., Hell, K. G. and Kerbauy, G. B. (1977). Tissue culture of *Stevia rebaudiana,* a sweetening plant. *Planta Medica,* 32, 115-117.

Hsing, Y. I., Su, W. F. and Chang, W. C. (1983). Accumulation of stevioside and rebaudioside A in callus culture of *Stevia rebaudiana* Bertoni. *Botanical Bulletin of Academia Sinica,* 24, 115-119.

Jayaraman, S., Manonharan, M. S., Illanchezian, S. (2008). *In vitro* antimicrobial activities of *S. rebaudiana* (Asteraceae) leaf. *Tropical Journal of Pharmaceutical Research,* 7 (4),1143-1149.

Khoda, H., Kasai, R., Yamasaki, K., Murakami, K. and Tanaka, O. (1976). New sweet diterpene glucosides from *Stevia rebaudiana. Phytochemistry,* 15, 981-983.

Kim, S. H. and Dubois, G. E. (1991). Natural high potency sweeteners. In Handbook of Sweeteners, S. Marie and J. R. Piggot (Eds), Avi, New York, pp. 116-185.

Kinghorn, A. D., Soejarto, D. D., Nanayakkara, N. P. D., Compadre, C. M., Makapugay, H. C. and Hovanec Brown, J. M. (1984). A phytochemical screening procedure for sweet ent-kaurene glycosides in the genus Stevia. *Journal of Natural Products,* 47, 439-444.

Kinghorn, A. D. and Soejarto D. D., Economic and Medicinal plant Research, Academic Pres, New York, USA, 1985, 1-51.

Kobayashi, M., Horikawa, S., Degrandi, I. H., Ueno, J. and Mitsuhashi, H. (1977). Dulcosides A and B, new diterpene glycosides from *Stevia rebaudiana. Phytochemistry,* 16, 1405-1408.

Lee, K. R., Park, J. R., Chol, B. S., Han, J. S., Oh, S. L., and Yamada, Y. (1982). Studies on the callus culture of Stevia as a new sweetening source and the formation of stevioside. Hanguk Sikpum Kwahakhoe Chi 14, 179-183 [Chemical Abstracts (1982) 97, 71003s]

Martelli, A., Frattini, C. and Chialva, F. (1985). Unusual essential oils with aromatic properties. Volatile components of *Stevia rebaudiana* Bertoni. *Flavour and Fragrance Journal,* 1, 3-7.

Matsuo, T., Kanamori, H. and Sakamoto, I. (1986). Non sweet glucoside in the leaves of Stevia rebaudiana. Hiroshima-ken Eisei Kenkyusho Kenkyu Hokoku 33, 25-29. [Chemical Abstracts (1987) 107, 93559o].

Melis, (1999). Effect of chronic administration of *Stevia rebaudiana* on fertility in rats. *Journal of Ethnopharmacology,* 67:157-161.

Metivier, J. and Viana, A. M. (1979). The effects of long and short day upon the growth of whole plants and the level of soluble proteins, sugars, and steviosides in the leaves of *Stevia rebaudiana* Bert. *Journal of Experimental Botany,* 30, 1211-1222.

Mizukami, H., Shiba, K., Santoshi, I. and Ohashi, H. (1983). Effects of temperature on growth and stevioside formation of *Stevia rebaudiana* Bertoni. *Shoyakugaku Zasshi,* 37, 175-179.

Nabeta, K., Kasai, T. and Sugisawa, H. (1976). Phytosterol from the callus of *Stevia rebaudiana* Bertoni. *Agricultural and Biological Chemistry,* 40, 2013-2014.

Oshima, Y., Saito, J. I. and Hikino, H. (1986). Sterebins A, B, C and D, bisnorditerpenoids of *Stevia rebaudiana* leaves. *Tetrahedron,* 42, 6443-6446.

Oshima, Y., Saito, J. I. and Hikino, H. (1988). Sterebins E., F., G. and H., diterpenoids of *Stevia rebaudiana* leaves. *Phytochemistry,* 27, 624-626.

Prakash, I., and Chaturvedula, V. S. P. (2013). Additional Minor Diterpene Glycosides from *Stevia rebaudiana* Bertoni. *Molecules,* 18, 13510-13519.

Prakash, I., Markosyan, A. and Bunders, C. (2014). Development of Next Generation Stevia Sweetener: Rebaudioside M. *Foods,* 3, 162-175.

Puri, M., Sharma, D., Barrow, Colin J. and Tiwary, A. K. (2012). Optimization of novel method for the extraction of steviosides from *Stevia rebaudiana* leaves. *Food chemistry,* 132, 3-112, 11130.

Rajbhandari, A. and Roberts, M. F. (1983). The flavonoids of *Stevia rebaudiana. Journal of Natural Products,* 46, 194-195.

Randi, A. M. and Felippe, G. M. (1981). Substances promoting root growth from the achenes of *Stevia rebaudiana. Revista Brasileira Botanica,* 4, 49-51. (Chemical Abstracts (1982)97, 52510 p).

Rao, A. B., Prasad E., Roopa G., Sridhar S., and Ravikumar, Y. V. L. (2012). Simple extraction and membrane purification process in isolation of steviosides with improved organoleptic activity. *Advances in Bioscience and Biotechnology,* 3, 327-335.

Robinson, H. and King, R. M. (1977).Eupatorieae-systemic review. In the Biology and chemistry of the Compositae. V. H. Heywood, J. B. Harborne and B. L. Turner (Eds.), Academic Press, London, 1, 437-485.

Sakamoto, I., Yamasaki, K. and Tanaka, O. (1977a). Application of ^{13}C NMR spectroscopy to chemistry of natural glycosides: rebaudiosides-C, a new sweet diterpene glycosides of *Stevia rebaudiana*. *Chemical and Pharmaceutical Bulletin*, 25, 844-846.

Sakamoto, I., Yamasaki, K. and Tanaka, O. (1977b). Application of ^{13}C NMR spectroscopy to chemistry of plant glycosides: rebaudiosides-D and E, a new sweet diterpene glycosides of *Stevia rebaudiana*. *Chemical and Pharmaceutical Bulletin*, 25, 3437-3439.

Samulsson and Gunnar, Drugs of National Origin, Swedish Pharmaceutical Press, Stockholm, Swedam, 1992.

Segura-Campos, M., Barbosa-Martín E., Matus-Basto, A., Cabrera-Amaro D., Murguía-Olmedo, M., Moguel-Ordoñez, Y. and Betancur-Ancona, D. (2014). Comparison of Chemical and Functional Properties of *Stevia rebaudiana* (Bertoni) Varieties cultivated in Mexican Southeast. *American Journal of Plant Sciences*, 5, 286-293.

Sholichin, M., Yamasaki, K., Miyama, R., Yahara, S. and Tanaka, O. (1980). Labdane types diterpenes from *Stevia rebaudiana*. *Phytochemistry*, 19, 326-327.

Soejarto, D. D., Kinghorn D. D. and Farnsworth N. R. (1982). Potential sweetening agents of plant origin. III. Organoleptic evaluation of *Stevia* leaf herbarium samples for sweetness. *Journal of Natural Products*, 45(5), 590-9.

Stevia The genus Stevia, 2002 edited by A. Douglas Kinghorn Department of Medicinal Chemistry and Pharmacognosy University of Illinois at Chicago USA @2002 Taylor and Francis pg no.68-70.

Stevia: Prospects as an Emerging Natural Sweetner. WHO INDIA 2007.

Suzuki, H., Ikeda, T., Matsumoto, T. and Noguchi, M. (1976). Isolation and identification of rutin from cultured cells of *Stevia rebaudiana* Bertoni. *Agricultural and Biological Chemistry*, 40, 819-820.

Swanson, S. M., Mahady, G. B. and Beecher, C. W. W. (1992). Stevioside biosynthesis of callus, root, shoot and rooted-shoot cultures *in vitro*. *Plant Cell Tissue and Organ Culture*, 28, 151-157.

Tamura, Y., Nakamura, S., Fukui, H. and Tabata, M. (1984). Comparison of Stevia plants grown from seeds, cuttings and stem tip cultures for growth and sweet diterpene glycosides. *Plant Cell Reports*, 3, 180-182.

Tanaka, O. (1982). Steviol-glycosides: new natural sweeteners. *Trends in Analytical Chemistry* 1, 246-248.

Viana, A. M., and Metiver, J. (1980). Changes in the levels of total soluble proteins and sugars during leaf ontogeny in *Stevia rebaudiana* Bert. *Annals of Botany*, 45,469-474.

Wada, E. R., Shibata R. and T. Torii. (1981). Nitrogen-15 abundance in Antarctica: Origin of soil nitrogen and ecological implications. *Nature, 292*, 327-329.

Yamazaki, K., Flores, H. E., Shimomura, K. and Yoshihira, K. (1991). Examination of steviol glucosides production by hairy root and shoot culture of *Stevia rebaudiana*. *Journal of Natural Products, 54*, 986-992.

In: Stevia rebaudiana
Editors: D. Betancur and M. Segura

ISBN: 978-1-63463-335-2
© 2015 Nova Science Publishers, Inc.

Chapter 4

ANTI-DIABETIC ACTIVITY OF *STEVIA REBAUDIANA* BERTONI AND ITS RELATIONSHIP WITH ITS ANTIOXIDANT PROPERTIES

J. C. Ruiz-Ruiz[1], Y. B. Moguel-Ordoñez[2], D. L. Cabrera-Amaro[2] and M. R. Segura-Campos[3]
[1]Departamento de Ingeniería Química-Bioquímica,
Instituto Tecnológico de Mérida, Mérida, Yucatán, México
[2]Instituto Nacional de Investigaciones Forestales, Agrícolas y Pecuarias,
Mocochá, Yucatán, México
[3]Facultad de Ingeniería Química, Universidad Autónoma de Yucatán.
Periférico Norte, Mérida, Yucatán, México

ABSTRACT

Recently, the treatment of diabetes mainly has involved a sustained reduction in hyperglycemia by the use of biguanides, thiazolidinediones, and sulfonylureas D-phenylalanine and α-glucosidase inhibitors in addition to insulin. However, due to unwanted side effects, the efficacies of these compounds are debatable, and there is a demand for new compounds for the treatment of diabetes. Furthermore, studies have specified that a hyperglycemia-induced overproduction of superoxide

* Phone: 52 999 946-09-56, Fax. 52 999 946-09-94. E-mail: maira.segura@uady.mx.

appears to be the major event in the development of complications of diabetes. Superoxide overproduction is associated with increased generation of nitric oxide, which results in formation of the strong oxidant peroxynitrite, and by polymerase activation, which, in turn, further initiates the pathways implicated in the development of diabetes-related complications. Hence, plants have been suggested as a rich, as yet unexplored source of potentially useful anti-diabetic drugs. Many traditional plant treatments for diabetes are used throughout the world. Plant drugs and herbal formulations are frequently considered to be less toxic and free from side effects than synthetic ones. Ethnobotanical studies have identified more than 1,200 species of plants with hypoglycemic activity throughout the world. Prior to the discovery of insulin, dietary measures and the traditional medicines derived from plants were the major forms of treatment. Among the anti-diabetic plants, *Stevia* (*Stevia rebaudiana* Bertoni), which is an herbaceous perennial plant native to subtropical and tropical rainforest areas of the South America, is one of the most efficacious plants. In this chapter, the antioxidant properties of the *Stevia rebaudiana* plant were explored in further depth. Understanding these factors and their mechanisms may be essential to comprehending the anti-diabetic effects of *Stevia rebaudiana* Bertoni in *in-vivo* models.

INTRODUCTION

Diabetes mellitus is a metabolic disorder initially characterized by a loss of glucose homeostasis with disturbances of carbohydrate, fat, and protein metabolism resulting from defects in insulin secretion, insulin action, or both (Barcelo and Rajpathak, 2001). Without enough insulin, the cells of the body cannot absorb sufficient glucose from the blood; hence, blood-glucose levels increase, which is termed hyperglycemia. If the glucose level in the blood remains high during a long period of time, this can result in long-term damage to organs, such as the kidneys, liver, eyes, nerves, heart, and blood vessels. Complications in some of these organs can lead to death (Pari and Saravanan, 2004). The pancreas plays a primary role in the metabolism of glucose by secreting the hormones insulin and glucagon (Figure 1). The islets of Langerhans secrete insulin and glucagon directly into the blood. Insulin is a protein that is essential for proper regulation of glucose and for maintenance of proper blood-glucose levels. Glucagon is a hormone that opposes the action of insulin. It is secreted when blood glucose level falls. It increases blood glucose concentration partly by breaking down stored glycogen in the liver by a pathway known as glycogenolysis. Gluconeogenesis is the production of glucose in the liver from non-carbohydrate precursors such as glycogenic amino acids (Pari and Saravanan, 2004).

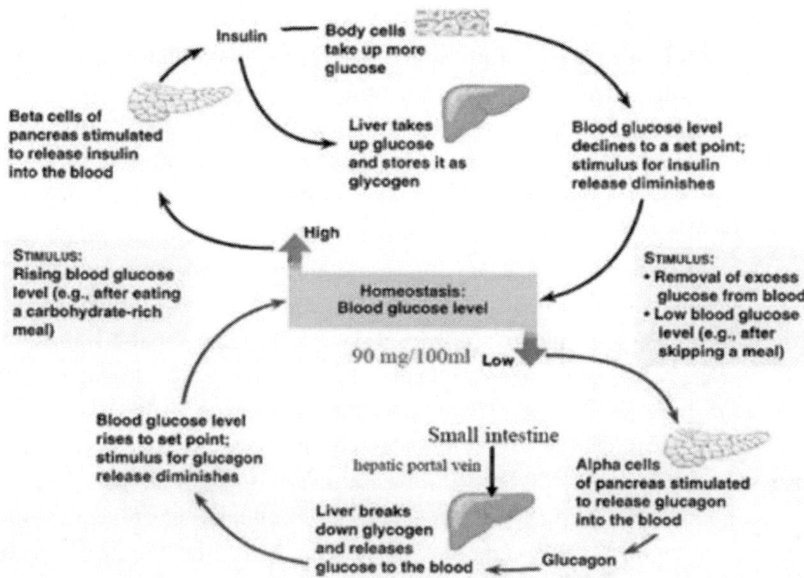

Figure 1. The role of pancreas in the body (Worthley, 2003).

The World Health Organization (WHO) classification of diabetes introduced in 1980 and revised in 1985 was based on clinical characteristics. The two most common types of diabetes were insulin-dependent diabetes mellitus (IDDM), or type I, and non-insulin-dependent diabetes mellitus (NIDDM), or type II. The WHO classification also recognized malnutrition-related diabetes mellitus and gestational diabetes. Malnutrition-related diabetes was omitted from the new classification because its etiology is uncertain, and it is unclear whether it is a separate type of diabetes (Holt, 2004; Tiwari and Rao, 2002). Type I diabetes mellitus is a result of cellular-mediated autoimmune destruction of the insulin secreting β-cells of the pancreas, which results in an absolute deficiency of insulin for the body. Patients are more prone to ketoacidosis. It occurs in children and young adults, usually before 40 years of age, although disease onset can occur at any age. The patient with type I diabetes must rely on insulin medication for survival. It may account for 5–10% of all diagnosed cases of diabetes. Autoimmune, genetic, and environmental factors are the major risk factors for type I diabetes

(Cavallerano and Cooppan, 2002; Trachtenbarg, 2005). Type II diabetes mellitus is characterized by a decreased ability of insulin to stimulate glucose uptake in peripheral tissues, insulin resistance, and the inability of the pancreatic β-cells to secrete insulin adequately (β-cell failure). Insulin resistance is defined as a state in a cell, tissue, system, or body for which levels of insulin needed to produce a quantitatively normal response are greater than normal. It is claimed that insulin has diverse effects and actions, depending on the different types of cells and tissues it is reacting to in the body. Insulin resistance is also closely related to hyperinsulinemia, though high blood-glucose is observed in the former while high insulin is observed in the latter. Insulin resistance also occurs in clinical settings such as pregnancy, cancer cachexia, obesity, starvation, burn trauma, sepsis, and as an outcome of several experimental treatments, both *in vivo* and *in vitro* (Houstis et al., 2006). Insulin-mediated glucose disposal is essentially impaired in most identified cases, as glucose levels are the main feedback signal for compensatory hyperinsulinemia (Pérez and Medina-Gómez, 2011). The major sites of insulin resistance in type II diabetes are the liver, skeletal muscle, and adipose tissue (Ostenson, 2001; White et al., 2003). Both defects (insulin resistance and β-cell failure) are caused by a combination of genetic and environmental factors. Environmental factors such as lifestyle habits (i.e., physical inactivity and poor dietary intake), obesity, and toxins may act as initiating factors or progression factors for type II diabetes. The genetic factors are still poorly understood (Uusitupa, 2002). Type II diabetes is increasingly being diagnosed at any age nowadays, and it accounts for 90–95% of all diagnosed cases of diabetes. It is associated with old age, obesity, family history of diabetes, impaired glucose metabolism, physical inactivity, and race/ethnicity (Holt, 2004; Li et al., 2004).

Excessively high levels of free radicals cause damage to cellular proteins, membrane lipids, and nucleic acids, and eventually cell death can occur. Various mechanisms have been suggested to contribute to the formation of these reactive free-oxygen radicals. Glucose oxidation is believed to be the main source of free radicals (Figure 2). In its enediol form, glucose is oxidized in a transition-metal dependent reaction to an enediol radical anion that is converted into reactive ketoaldehydes and to superoxide anion radicals. The superoxide anion radicals undergo dismutation to hydrogen peroxide, which, if not degraded by catalase or glutathione peroxidase and in the presence of transition metals, can lead to production of extremely reactive hydroxyl radicals (Maritim et al., 2003).

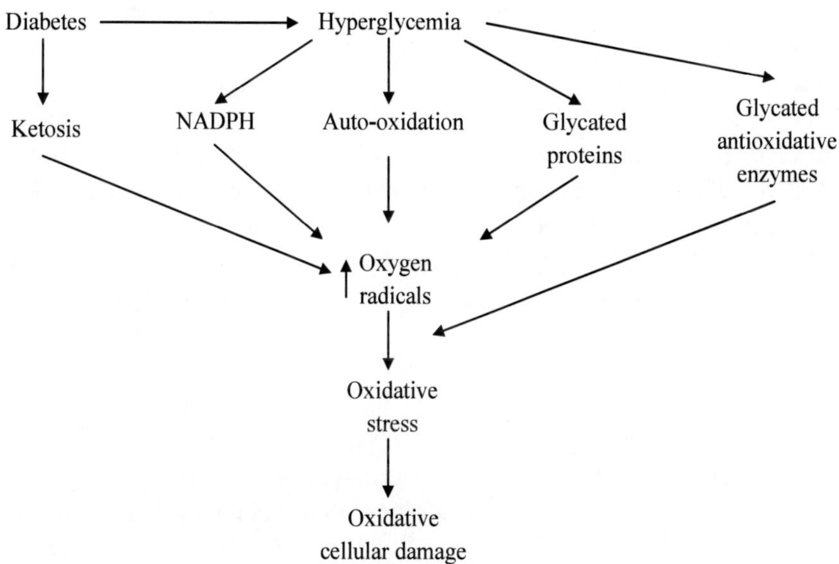

Figure 2. Oxidative stress generated by diabetes (Jain, 2000).

Current studies have specified that a hyperglycemia-induced overproduction of superoxide appears to be the major event in the development of complications of diabetes. Superoxide overproduction is associated with increased generation of nitric oxide, which results in formation of the strong oxidant peroxynitrite, and by poly (adenosine diphosphate-ribose) polymerase activation, which, in turn, further initiates the pathways implicated in the development of diabetes-related complications. In addition, this process causes a severe endothelial dysfunction and the initiation of inflammation in blood vessels of individuals with diabetes, and these aspects contribute to the development of complications of diabetes like cardiovascular problems such as coronary heart disease, peripheral arterial disease, hypertension, stroke, cardiomyopathy, nephropathy, and retinopathy (Ceriello, 2006).

PREVALENCE, INCIDENCE, AND PHARMACOLOGICAL MANAGEMENT OF DIABETES MELLITUS

The prevalence of diabetes mellitus is increasing with aging of the population and lifestyle changes associated with rapid urbanization and westernization. The disease is found in all parts of the world and is rapidly increasing in coverage (Kamalakkanan and Prince, 2003). Globally, the prevalence of diabetes, without type distinction, was estimated to be 4% in 1995. According to the WHO, it is estimated that 3% of the world's population have diabetes, and the prevalence is expected to double by the year 2025 to 6.3% (Andrade-Cetto and Heinrich, 2005). There will be a 42% increase from 51–72 million in the developed countries and a 170% increase from 84–228 million in the developing countries. Thus, by the year 2025, more than 75% of all people with diabetes will be in the developing countries, compared to 62% in 1995 (Ramachandran et al., 2002). The reasons behind this projected increase in prevalence rate are due to urbanization, westernization, lifestyle changes associated with those processes, increase in life expectancy at birth, physical inactivity, obesity, and possibly a genetic predisposition (Wild et al., 2004). Age, ethnic, regional, and racial differences also have been found to play a role in the diabetic incidence in heterogeneous populations within the same area (Alberti et al., 2007).

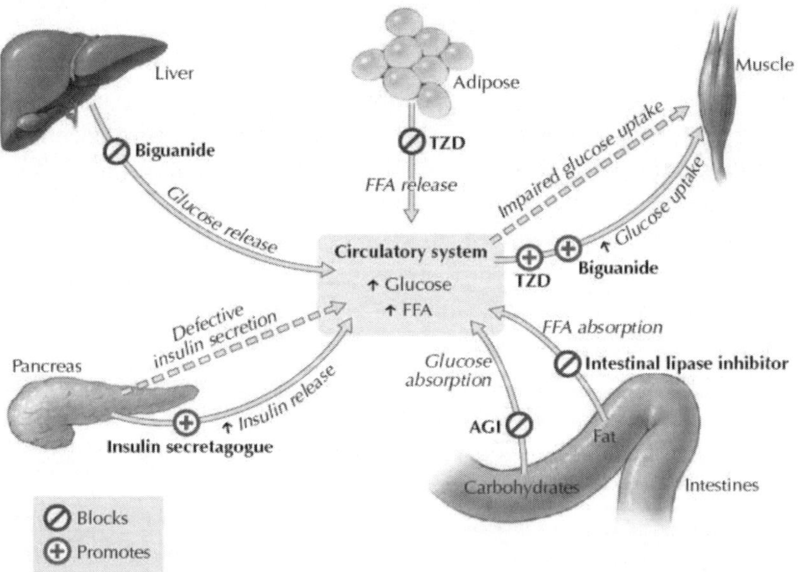

Figure 3. Major target organs and mechanism of actions of orally administered antihyperglycemic agents in type II diabetes mellitus (Cheng and Fantus, 2005). TZD = thiazolidinedione; FFA = free fatty acid; AGI = α-glucosidase inhibitor.

Diet, exercise, modern drugs such as insulin, and oral administration of hypoglycaemic drugs such as sulfonylureas and biguanides manage the pathogenesis of diabetes mellitus. Insulin plays a key role in glucose homeostasis alongside the counter-regulatory hormone glucagon, which raises serum glucose. Glucose transport proteins (GLUT 1–5) are essential for glucose uptake into cells. In individuals with type II diabetes, a common sequence of therapy starts with diet treatment and exercise followed by oral antihyperglycemic agents. There is no drug available in the market for a permanent cure of diabetes. There are many oral hypoglycemic agents that are available in the market like metformin hydrochloride, glimepiride, and glibenclamide. They are effective but have side effects including lactic acidosis, metallic taste, vitamin B12 deficiency, and others. Oral agents acting as indicated are used in type II diabetic patients who fail to meet glycemic goals with medical nutrition therapy and exercise (Figure 3).

Insulin has been used successfully in insulin-dependent diabetes mellitus since 1992. However, it has limitations such as it cannot be given orally; daily administration through injection is troublesome. Besides these, hypoglycemic reactions may occur in any diabetic patient treated with it. In patients who are taking insulin for long periods, insulin resistance may also occur (Larner, 2001). Traditionally, plants and its infusions are also used for the treatment of diabetes throughout the world (Koski, 2006; Pari and Saravanan, 2004). Management of diabetes without any side effects is still a challenge for the medical system. This leads to an increasing search for improved anti-diabetic drugs.

THE IMPORTANCE OF *STEVIA REBAUDIANA* ON DIABETES TREATMENT

Medicinal plants, since time immemorial, have been used in virtually all cultures as a source of medicine. It has been estimated that about 80–85% of populations both in developed and developing countries rely on traditional medicine for their primarily healthcare needs, and it is assumed that a major part of traditional therapy involves the use of plant extracts or their active principles. The active principles of many plant species are isolated for direct use as drugs, lead compounds, or pharmacological agents. Ethnobotanical information indicates that more than 500 plants are used as traditional remedies for diabetes mellitus treatment (Suanarunsawata et al., 2004), and more than 1200 species of plants exhibit hypoglycemic activity (Okyar et al., 2001). For diabetes treatment, before the discovery of insulin by Banting and Best in 1922, the only options were those based on traditional practices (Ribnicky et al., 2006). Today, metformin, which is derived from the medicinal plant *Galega officinalis*, is the only ethical drug approved for the treatment of non-insulin dependent diabetes mellitus patients (Eshrat, 2002). Among those plants used traditionally for the treatment of diabetic complications is *Stevia rebaudiana* Bertoni (Li et al., 2004).

Stevia, one of the 950 genera of the Asteraceae family, is a genus of more than 200 species. Members of *Stevia* comprise mostly of herbs but also shrubs and trees. Originally, it is said to be native to subtropical South America (Paraguay and Brazil) and Central America but now is found over a wide range of areas of 500–3500 m altitude, 1500–1800 mm rainfall, and -6–43 °C temperatures (De Oliveira et al., 2004). *Stevia rebaudiana* usually grows in semi-dry mountainous terrains. Its habitats range from grasslands, forested mountain slopes, and conifer forests to sub-alpine vegetation. It is an herb 80–180 cm tall with a lifespan of 3–5 years. It grows best in soil that is well drained but with reasonable water-holding capacity and preferably with pH 5–7; alkaline soil should be avoided (Uddin et al., 2006). For centuries, the Guarani Indians in Paraguay and Brazil used *Stevia* species, primarily *S. rebaudiana*, as a non-calorie sweetener (250–300 times sweeter than sucrose at 0.4% solution) in medicinal green teas for treating heartburn and other ailments (Vanek et al., 2001). Although there are more than 200 species of the genus *Stevia*, *S. rebaudiana* gives the sweetest essence (Savita et al., 2004).

ANTI-DIABETIC EFFECT OF STEVIOL GLYCOSIDES

The natives of Paraguay and Brazil have used *Stevia rebaudiana* and its major steviol glycoside, stevioside, as sugar substitute and medical treatment for centuries. Apart from its sweetness, *S. rebaudiana* and stevioside also have been reported to induce various physiological effects such as cardiovascular, renal, microbiological, and endocrine effects (Chan et al., 2000). Suanarunsawata et al. (2004) evaluated the glycemic effect and the mechanism of action of both stevioside (0.25 g/kg body weight) and *S. rebaudiana* extract (4.66 g/kg body weight) in streptozotocin-induced diabetic rats for eight weeks. These authors observed that plasma glucagon (PG) level was slightly but significantly raised in normal rats fed with stevioside since the third week.

In contrast, *S. rebaudiana* had no significant effect on the PG in normal rats. Stevioside had no effect on the PG in diabetic rats, whereas *S. rebaudiana* significantly reduced the PG in diabetic rats from the second week until the end of experiment. The mechanism of glycemic action of stevioside and *S. rebaudiana* was correlated with the alteration of serum insulin levels. There were no changes in normal rats fed with either stevioside or *S. rebaudiana*. In contrary, stevioside and *S. rebaudiana* improved the serum insulin in diabetic rats. Suanarunsawata et al. (2004) concluded that stevioside potentiated insulin release in diabetic rats, a reduction of insulin-induced glucose uptake was apparent as well; thereby no anti-hyperglycemic effect was apparent. *S. rebaudiana* decreased PG in diabetic rats by the mechanism of an improvement of insulin and suppression of glucagon level. Stevioside containing in *S. rebaudiana* may indirectly contribute to anti-hyperglycemic action of *S. rebaudiana* in diabetic rats via its effect to potentiate insulin release.

In other study, Dutta et al. (2010) compared the hypoglycemic activity of *Stevia* leaves and metformin hydrochloride and also compared the short-term adverse effects of them on body weight. The author concluded that the continuous treatment with *Stevia* produced a significant reduction of the blood-glucose level and that metformin also significantly reduced (p<0.001) blood level in streptozotocin-induced diabetic rats. Among three doses of 10%-aqueous extracts of *Stevia* (2, 5, and 10 mL/kg body weight), 10 mL/kg body weight was more potent. The aqueous extract of *Stevia* leaves decreased the body weight insignificantly at 2 mL/kg body weight but significantly at 5 and 10 mL/kg body weight. Metformin hydrochloride decreased the body weight but not significantly (p>0.05). Therefore, it is evident that *Stevia* has hypoglycemic activity like metformin hydrochloride, but it decreases body weight. Klepser and Kelly (1997) established that metformin hydrochloride lowers blood-glucose levels by stimulating glucose uptake of skeletal muscle. According to the results obtained by Dutta et al. (2010) and Suanarunsawata et al. (2004), the extract of *S. rebaudiana* could exhibit the same mechanism of action as metformin.

Mishra (2011) performed an analysis of anti-diabetic activity of *Stevia rebaudiana* extract on diabetic patients. Fifteen patients who were vegetarian were introduced; nine patients were women, six patients were men, and the age group was 35–60 years. During the experimental period, the patients did not use drugs. *Stevia* leaf powder (0.5–1 g) was used in place of sugar in tea/coffee and milk intake. The feeding trial was 45 days. During the first 15 days, patients were given medicine, and their fasting and post-prandial diabetic levels were measured. Then, for next 15 days, they were not given any medicine while under normal diet, and their prandial and fasting glucose levels were measured. For 15 more days, they were given *Stevia* leaves three times a day, and their glucose levels were measured. Initially, for 15 days patients were taking medicine regularly as they were earlier under normal diet, and their fasting blood sugar (FBS) and post-prandial (PPBS) blood sugar levels were taken. The FBS was 153.54 and PPBS was 189.56 (means). For next 15 days, patients were asked to stop taking medicine under normal diet, and, under supervision of a physician, their FBS and PPBS were checked, which were 208.6 and 283 (means), respectively. There were significant differences in both FBS and PPBS that showed that while taking medicine, their FBS and PPBS levels were under control. For 15 more days, patients were given *Stevia* leaf powder three times, and FBS and PPBS levels were checked, which were 195.7 and 271.3 (means), respectively. The authors concluded that *S. rebaudiana* extract decreased FBS and PPBS. However, the results had no statistical significance, probably due to the sample size (15 patients) not being enough to show significant results.

ANTI-DIABETIC EFFECT MEDIATED BY ANTIOXIDANT COMPOUNDS

As a whole, researchers worldwide agree on the anti-diabetic effects of *Stevia*; but, they differ on how the effects contribute toward combating this metabolic disease. It is important to note that there are many steviol glycosides, which are compounds with multiple carbohydrate molecules, bound to a non-carbohydrate, aglycone moiety (steviol) that can be extracted from the *Stevia rebaudiana* Bertoni plant. The most commonly extracted glycosides are stevioside, rebaudioside A, rebaudioside C, and dulcoside, among many others available (Lemus-Mondaca et al., 2012). Some assert that *Stevia*'s utility is due to its antioxidant properties; this is supported by analysis of the phenols that may be extracted from the plant. *Stevia* has a large overall proportion of phenols, up to 91 mg/g; it is proposed that these constituents extracted from the leaves are the major agents contributing toward the anti-hyperglycemic activities exerted by the plant. This is further supported by the fact that the leaves have a greater ability to scavenge free radicals and prevent lipid peroxidation than controls such as butylated hydroxytoluene, butylated hydroxyanisole, and tertiary butyl hydroxyquinone (Shivanna et al., 2012).

Such findings concur with the results of other studies of Type I diabetes, modeled by streptozotocin-induced diabetic rats, in which phenolic compounds prevented several diabetic complications. In addition, Shivanna et al. (2012) observed a significant decrease (about 30%) in peroxidation in the livers of *Stevia*-pre-fed rats compared to those of their control groups. This is a good indicator of reduction in the progression of diabetic complications, as diabetic tissue damage is commonly linked to the peroxidation of lipids. Likewise, the condition of hyperglycemia, which increases the production of reactive oxygen species (ROS) in the tissues due to high blood-glucose levels, makes the tissues susceptible to oxidation (Das et al., 2000).

Jahan et al. (2010) performed four complementary test systems: DPPH free-radical scavenging, reducing power, total phenolics, and total flavonoids concentration determinations in order to study the antioxidant capacities of different organic and aqueous soluble materials of *Stevia rebaudiana* leaves. The IC_{50} values of the 80%-ethanol extract and water-soluble fractions were found to be 8.02 and 43.81 µg/mL, respectively. In the same test, the standard compound (ascorbic acid) exhibited an IC_{50} value 4.21 µg/mL. In reducing power tests, the maximum absorbance at 700 nm for the 80%-ethanol extract and water-soluble fractions was found to be up to 1.071 and 1.555, respectively, compared to the absorbance of ascorbic acid as a standard (1.374). Total phenolic concentrations in the 80%-ethanol extract and water soluble fractions were 65.21 and 41.49 mg of gallic acid equivalent/g of dry extract, respectively. Total flavonoid concentrations in the 80%-ethanol extract and water soluble fractions were 125.64 and 101.45 mg of quercetin equivalent/g of dry extract, respectively. The results obtained by these authors revealed that the 80%-ethanol extract exhibited the most significant antioxidant activity. In another study, Shivanna et al. (2012) identified by mass spectrometry the major phenolic compounds in *Stevia* powder extracted with ethanol: dicaffeoylquinic acid, chlorogenic acid, quercetin 3-O-xyloside, apigenin-7-O-glucoside, 3,4-dimethoxycinnamic acid, luteolin-7-O-rutinoside, and caffeic acid. Previous studies had noted the importance of hypoglycemic components of *Stevia* is due to the rebaudioside A and stevioside concentrations in leaves (Wheeler et al., 2008). However, Shivanna et al. (2012) did not identify these compounds in the polyphenol extract they studied. This explains the involvement of polyphenolic compounds in preventing diabetic and its complications caused by the streptozotocin model (Bennet and Pegg, 2004; Imaeda et al., 2002; Traverso et al., 2004). The same authors compared the antioxidant properties of *Stevia* leaves with other commercial antioxidants such as butylated hydroxyanisole, butylated hydroxytoluene, and tertiary butyl hydroquinone. *Stevia* leaves exhibited better antioxidant properties such as free-radical scavenging (EC_{50} = 2.6 µg) and inhibition of lipid peroxidation (EC_{50} = 10.6 µg). From the results observed by Jahan et al. (2010) and Shivanna et al. (2012) of antioxidant study following different *in-vitro* methods, it can be concluded that ethanolic and water extracts of *S. rebaudiana* leaves possess significant antioxidant activity that could be related with its abrogating effect on insulin resistance.

The relationship between the antioxidant activities of *S. rebaudiana* and its abrogating effect on insulin resistance was evaluated by Shivanna et al. (2012). These authors evaluated the oxidative stress in livers of streptozotocin-induced diabetic rats (60 mg/kg body weight) by measuring the thiobarbituric acid reaction substances (TBARS), conjugated dienes, and hydroperoxides, which are products of lipid peroxidation. Peroxidation was reduced significantly in rats previously fed with *Stevia* leaf powder and polyphenols extracted (4.0 g of leaf powder or phenolic extract in 96 g of dry diet) 25% and 30% in the liver compared to the diabetic group. Since high blood glucose is susceptible to oxidation, hyperglycemia causes high ROS production and, in turn, leads to high malondialdehyde levels in tissues (Das et al., 2000). However, the increment of lipid peroxidation has been found to be involved in observed tissue damage in diabetes (Ahmed et al., 2006; Can et al., 2004; Mahboob et al., 2005). Shivanna et al. (2012) observed that streptozotocin administration resulted in a significant decrease in vitamin-C content by 27%, an increase in vitamin-E levels by 130%, and an increase in glutathione or the glutathione/glutathione disulfide ratio by 45% compared to rats pre-fed with *Stevia* leaf powder and polyphenols extracted that did not alter the levels of these antioxidants in plasma. Streptozotocin administration also resulted in significant reduction of hepatic antioxidant enzymes super oxide dismutase (SOD) and catalase (CAT) by 50%. When rats were previously fed with *Stevia* leaf powder and polyphenols extracted the enzyme activity was restored to normal. According to these authors, pre-treatment with *Stevia* stimulated SOD and CAT to reverse oxidative damage. The results indicate that *Stevia* showed significant protection against the oxidative damage in livers of diabetic rats induced by streptozotocin.

Singh et al. (2013) evaluated the antihyperglycemic and antioxidative properties of methanolic extracts of *Stevia rebaudiana* leaves in alloxan-induced diabetic mice. Alloxan is a specific toxin that causes massive destruction of the pancreatic β-cells, provoking a state of primary deficiency of insulin without affecting other islet types, thus creating a hyperglycemic condition. However, after continuous treatment of diabetic mice with *S.*

rebaudiana for 21 days, the authors observed a considerable reduction of 39.84% (193.0 down to 116 mg/dL) in the sugar level. In the liver, kidney, and pancreas, the authors estimated changes in SOD, glutathione peroxidase (GSH-Px), reduced glutathione (GSH), and TBARS. According to Singh et al. (2013), treatment with leaf extract of *S. rebaudiana* did not show a considerable improvement in the SOD level of pancreatic and renal tissues. However, hepatic level was normalized in the case of SOD. Moreover, treatment with the leaf extract caused a significant increase in the level of GPx, suggesting its compensatory role in reducing hydrogen peroxide produced, thus diminishing the toxic effects of free radicals produced by it in various secondary reactions. After the extract treatment for 21 days, a significant increase in the GSH level was observed, which indicates that the leaf extract of *S. rebaudiana* has the potential to increase the biosynthesis of GSH, thereby reducing the oxidative stress. According to Sabu and Kuttan (2002), the diabetogenic effect of alloxan is due to excess production of ROS, leading to cytotoxicity in pancreatic β-cells, which reduces the synthesis and the release of insulin while affecting organs such as the liver, kidney, pancreas, and hematopoietic system. Hyperglycemia generates ROS, which, in turn, cause lipid peroxidation and thus membrane damage (Hunt et al., 1988). In their study, Singh et al. (2013), using alloxan induction reported an increase in the level of TBARS concentrations in diabetic mice. However, the treatment of methanolic leaf extract significantly reduced the level of TBARS. The multiple effects observed by Singh et al. (2013) could be due to the presence of some biomolecules present in the plant extract that may have stimulated the β-cells of Langerhans to release insulin, leading to the improvement in the carbohydrate-metabolizing enzymes and thus establishing normal blood-glucose levels (Hossain et al., 2011). According to diverse studies, extracts of leaves of *S. rebaudiana* exhibit the presence of various phytochemicals such as alkaloids, flavonoids, phenols, steroids, and tannins (Preethi et al., 2011; Shivanna et al., 2012). Most of these compounds could have a marked antioxidant effect on cells, tissues, or organs.

CONCLUSION

Streptozotocin- or alloxan-induced diabetes in rats represent well-established animal models of type I, insulin-dependent diabetes mellitus. Increased production of high levels of free-oxygen radicals has been linked to glucose oxidation and non-enzymatic glycation of proteins, which contribute to the development of diabetic complications. Protective effects of exogenously administered antioxidants such as plants extracts have been extensively studied in animal models within recent years, thus providing some insight into the relationship between free radicals, diabetes, and its complications. Exhaustive work has been done on *Stevia rebaudiana* Bertoni leaf powders as well as its organic and aqueous extracts, but there is still need for research to understand the mechanisms and compounds involved in their anti-diabetic effects. It is necessary to establish the pharmacological aspects of steviol glycosides and phytochemicals of *S. rebaudiana*, its adverse effects, absorption, distribution, metabolism, and excretion in order to promote its use as a therapeutic agent or nutraceutical ingredient in functional foods.

REFERENCES

Ahmed, F.N., Naqvi, F.N., Shafiq, F. (2006). Lipid peroxidation and serum antioxidant enzymes in patients with type 2 diabetes mellitus. *Annals of the New York Academy of Science, 1084,* 481-489.

Alberti, K.G.M., Zimmet, P. and Shaw, J. (2007). International Diabetes Federation: a consensus on Type 2 diabetes prevention. *Diabetic Medicine, 24,* 451-463.

Andrade-Cetto, A. and Heinrich, M. (2005). Mexican plants with hypoglycemic effect used in the treatment of diabetes. *Journal of Ethnopharmacology, 99,* 325-348.

Barcelo, A. and Rajpathak, S. (2001). Incidence and prevalence of diabetes mellitus in the Americas. *American Journal of Public Health, 10,* 300-308.

Bennett, R.A. and Pegg, A.E. (2004). Alkylation of DNA in rat tissues following administration of streptozotocin. *Cancer Research, 4,* 2786-2790.

Can, A., Akev, N., Ozsoy, N., Bolkent, S., Arda, B.P., Yanardag, R. (2004). Effect of *Aloe vera* leaf gel and pulp extracts on the liver in type-II diabetic rat models. *Biological and Pharmaceutical Bulletin, 27,* 694-698.

Cavallerano, J.O.D. and Cooppan, R. (2002). Optometric clinical practice guideline care of the patient with diabetes mellitus. American Optometric Association 243 N. Lindbergh Blvd., St. Louis, MO. 63141-7881, 3rd ed., U.S.A.

Ceriello, A. (2006). Oxidative stress and diabetes-associated complications. *Endocrinology Practice, 12(1),* 60-62.

Chan, P., Tomlinson, B., Chen, Y.J., Liu, J.C., Hsieh, M.H, Cheng, J.T. (2000). A double-blind placebo controlled study of the effectiveness and tolerability of oral stevioside in human hypertension. *British Journal of Clinical Pharmacology, 50,* 215-220.

Cheng, A.Y.Y. and Fantus, I.G. (2005). Oral antihyperglycemic therapy for type 2 diabetes mellitus. *Canadian Medical Association Journal, 172,* 213-226.

Das, S., Snehlata, S.V., Das, N., Srivastava, L.M. (2000). Correlation between total antioxidant status and lipid peroidation in hypercholesterolemia. *Current science, 78,* 486-487.

De Oliveira, V.M., Forni-Martins, E.R., Magalhães, P.M., Alves, M.N. (2004). Chromosomal and morphological studies of diploid and polyploid cytotypes of *Stevia rebaudiana* Bertoni (Eupatorieae, Asteraceae). *Genetics and Molecular Biology, 27,* 215-222.

Dutta, P.K., Razu, M.M.T., Alam, M.K., Awal, M.A., Mostofa, M. (2010). Comparative efficacy of aqueous extract of Stevia (*Stevia rebaudiana* Bertoni) leaves and metformin hydrochloride (Comet$^®$) in streptozotocin induced diabetes mellitus in rats. *International Journal of Biology Research, 2(8),* 17-22.

Eshrat, H.M.A. (2002). Hypoglycemic, hypolipidemic and antioxidant properties of combination of curcumin from curcuma longa, linn, and partially purified product from *Abroma augusta*, Linn. In streptozotocin induced diabetes. *Indian Journal of Clinical Biochemistry, 17,* 33-43.

Holt, R.I.G. (2004). Diagnosis, epidemiology and pathogenesis of diabetes mellitus: an update for psychiatrists. *British Journal of Psychiatry, 184(47),* s55-s63.

Hossain, M.S., Alam, M.S., Asadujjaman, M., Islam, M.M., Rahman, M.A., Islam, M.A., Islam A. (2011). Antihyperglycemic and antihyperlipidemic effects of different fractions of *Stevia rebaudiana* leaves in alloxan induced diabetic rats. *International Journal of Pharmaceutical Sciences Research, 287,* 1722-1729.

Houstis, N., Rosen, E.D., Lander, E.S. (2006). Reactive oxygen species have a causal role in multiple forms of insulin resistance. *Nature, 440,* 944-948.

Hunt, J.V., Dean, R.T., Wolff, S.P. (1988). Hydroxyl radical production and autoxidative glycosylation. Glucose autoxidation as the cause of protein damage in the experimental glycation model of diabetes and aging. *Biochemical Journal, 256,* 205-212.

Imaeda, A., Kaneko, T., Aoki, T., Kondo, Y., Nagase, H. (2002). DNA damage and the effect of antioxidants in streptozotocin-treated mice. *Food and Chemical Toxicology, 40,* 979-987.

Jahan, I.A., Mostafa, M., Hossain, H., Nimmi, I., Sattar, A., Alim, A., Iqbal-Moeiz. S.M. (2010). Antioxidant activity of *Stevia rebaudiana* Bert. Leaves from Bangladesh. *Bangladesh Pharmaceutical Journal, 13(2),* 67-75.

Jain, S.K. (2000). The mechanism(s) of complications and benefits of vitamin e supplementation in diabetic patients. *Diabetologia Croatica, 29* (3).

Kamalakkanan, N., and Prince, P.S.M. (2003). Hypoglycaemic effect of water extracts of Aegle *marmelos* fruits in streptozotocin-diabetic rats. *Journal of Ethnopharmacology, 87,* 207-210.

Klepser, T.B. and Kelly, M.W. (1997). Metformin hydrochloride: an antihyperglycemic agent. *American Journal of Health-System Pharmacy, 54(8),* 893-903.

Koski, R.R. (2006). Practical review of oral antihyperglycemic agents for Type 2 diabetes mellitus. *The Diabetes Educator, 32,* 869-876.

Larner, J. (2001). Insulin and oral hypoglycemic drug, glucagon. In: Goodman and Gillman's the Pharmacological Basis of Therapeutics. Gilman A.G., Goodman L.S., Rail T.W. and Murad, F. (eds.), vol. 2, 10th ed. The MacMillan Publishing Co., New York, Chapter 61.

Lemus-Mondaca, R., Vega-Galvez, A., Zura-Bravo, L., Ah-Hen, K. (2012). Optimisation of novel method for the extraction of steviosides from *Stevia rebaudiana* leaves. *Food Chemistry, 132(17),* 1121-1132.

Li, W. L., Zheng, H.C., Bukuru, J., De Kimpe, N. (2004). Natural medicines used in the traditional Chinese medical system for therapy of diabetes mellitus. *Journal of Ethnopharmacology, 92,* 1-21.

Mahboob, M., Rahman, M.F., Grover, P. (2005). Serum lipid peroxidation and antioxidant enzyme levels in male and female diabetic patients. *Singapore Medical Journal, 46,* 322-324.

Maritim, A.C., Sanders, R.A., Watkins, J.B. (2003). Diabetes, Oxidative Stress, and Antioxidants: A Review. *Journal of Biochemical and Molecular Toxicology, 17(1),* 24-37.

Mishra, N. (2011). An Analysis of antidiabetic activity of *Stevia rebaudiana* extract on diabetic patient. *Journal of Natural Sciences Research, 1(3).*

Okyar, A., Matsuda, K., Sato, M., Yamagata, S. and T. Sato. (2001). Effect of *Aloe vera* leaves on blood glucose level in type I and type II diabetic rat Models. *Phytotherapy Research, 15(2),* 157-161.

Ostenson, C.G. (2001). The pathophysiology of type 2 diabetes mellitus: an overview. *Acta Physiology of Scandinavian, 171,* 241-247.

Pari, L. and Saravanan, R. (2004). Antidiabetic effect of diasulin, a herbal drug, on blood glucose, plasma insulin and hepatic enzymes of glucose metabolism in hyperglycaemic rats. *Diabetes, Obesity and Metabolism, 6,* 286-292.

Pérez, M.R. and Medina-Gómez, G. (2011). Obesidad, adipogénesis y resistencia a la insulina. *Endocrinológica y Nutrición, 58(7),* 360-369.

Preethi, D., Sridhar, T.M., Josthna, P., Naidu, C.V. (2011). Studies on antibacterial activity, phytochemical analysis of *Stevia rebaudiana* (Bert.). An important calorie free biosweetner. *Journal of Ecobiotechnology, 3(7),* 5-10.

Ramachandran, A., Snehalatha, C., Viswanathan, V. (2002). Burden of type 2 diabetes and its complications. The Indian scenario. *Current Science, 83,* 1471-1476.

Ribnicky, D.M., Poulev, A., Watford, M., Cefalu, W.T., Raskin, I. (2006). Antihyperglycemic activity of Tarralin, an ethanolic extract of *Artemisia dracunculus* L. *Phytomedicine, 13,* 550-557.

Sabu, M.C. and Kuttan, R. (2002). Anti-diabetic activity of medicinal plants and its relationship with their antioxidant property. *Journal of Ethnopharmacology, 81(29),* 155-160.

Savita, S.M., Sheela, K., Sunanda, S., Shankar, A.G., Ramakrishna, P., Sakey, S. (2004). Health implications of *Stevia rebaudiana. Journal of Human Ecology, 15,* 191-194.

Shivanna, N., Naika, M., Khanum, F., Kaul, V.K. (2012). Antioxidant, anti-diabetic and renal protective properties of *Stevia rebaudiana. Journal of Diabetes and Its Complications, 27(2),* 103-113.

Singh, S., Garg, V., DEEPAK Yadav, D. (2013). Antihyperglycemic and antioxidative ability of *Stevia rebaudiana* (Bertoni) leaves in diabetes induced mice. *International Journal of Pharmacy and Pharmaceutical Sciences, 5(2),* 297-302.

Suanarunsawata, T., Klongpanichapaka, S., Rungseesantivanona, S., Chaiyabutrb, N. (2004). Glycemic effect of stevioside and *Stevia rebaudiana* in streptozotocin-induced diabetic rats. *Eastern Journal of Medicine, 9,* 51-56.

Tiwari, A.K. and Rao, J.M. (2002). Diabetes mellitus and multiple therapeutic approaches of phytochemicals: Present status and future prospects. *Current Science, 83,* 30-38.

Trachtenbarg, D.E. (2005). Diabetic ketoacidosis. *American Family Physician,* 71: 1705-1714.

Traverso, N., Menini, S., Maineri, E. P., Patriarca, S., Odetti, P., Cottalasso, D. (2004). Malondialdehyde, a lipoperoxidation-derived aldehyde, can bring about secondary oxidative damage to proteins. *Journals of Gerontology Series A Biological Sciences and Medical Sciences, 59,* 890-895.

Uddin, M.S., Chowdhury, M.S.H., Khan, M.M.H., Uddin, M.B., Ahmed, R. and Baten, M.A. (2006). *In vitro* propagation of *Stevia rebaudiana* Bert in Bangladesh. *African Journal of Biotechnology, 5,* 1238-1240.

Uusitupa, M. (2002). Lifestyles matter in the prevention of Type 2 diabetes. *Diabetes care, 25,* 1650-1651.

Vanek, T., Nepovim, A., Valicek, P. (2001). Determination of Stevioside in plant material and fruit teas. *Journal of food composition and analysis, 14,* 383-388.

Wheeler, A., Boileau, A.C., Winkler, P.C., Compton, J.C., Prakash, I., Jiang, X. (2008). Pharmacokinetics of rebaudioside A and stevioside after single oral doses in healthy men. *Food and Chemical Toxicology, 46(7),* S54-S60.

White, J.R., Davis, S.N., Cooppan, R., Davidson, M.B., Mulcahy, K., Manko, G.A., Nelinson, D. (2003). Clarifying the role of insulin in Type 2 diabetes management. *Clinical Diabetes, 21,* 14-21.

Wild, S., Roglic, G., Green, A., Sicree, R., King, H. (2004). Global prevalence of diabetes estimates for the year 2000 and projections for 2030. *Diabetes Care, 27,* 1047-1053.

Worthley, L.I.G. (2003). The Australian short course on intensive care medicine, Handbook, Gillingham printers, South Australia, pp 31-55.

In: Stevia rebaudiana
Editors: D. Betancur and M. Segura

ISBN: 978-1-63463-335-2
© 2015 Nova Science Publishers, Inc.

Chapter 5

STEVIA REBAUDIANA BERTONI ANTIOXIDANT ACTIVITY AND ITS PRESERVATIVE POTENTIAL COMBINED WITH HIGH HYDROSTATIC PRESSURE

*María Nieves Criado, Clara Miracle Belda-Galbis,
Antonio Martínez and Dolores Rodrigo**

Instituto de Agroquímica y Tecnología de Alimentos (IATA-CSIC),
Carrer del Catedràtic Agustín Escardino Benlloch,
Paterna, València, Spain

ABSTRACT

In recent years, a possible alternative to complement the effectiveness of non-thermal technologies is to combine these technologies with the addition of preservatives of natural origin with antimicrobial and antioxidant properties. Given that the antioxidant properties of *Stevia rebaudiana* Bertoni (*Stevia*) have only been studied on the basis of chemical reactions and, moreover, no studies about its potential to enhance the effectiveness of non-thermal preservation technologies have been conducted to date, the antioxidant capacity of *Stevia*, determined as polyphenoloxidase (PPO) and peroxidase (POD) percentage of inhibition at 10 and 37°C, and also the preservative

---

*Corresponding author: lolesra@iata.csic.es

potential of *Stevia* in combination with high hydrostatic pressure were evaluated.

The results obtained show that *Stevia* can inhibit PPO and POD activities in an enzymatic extract obtained from a mixture of orange, mango and papaya. In addition, it was observed that a 2.5% (w/v) leaf infusion could enhance the antimicrobial and antioxidant potential of HHP because *Listeria monocytogenes* inactivation and inhibition of the oxidative enzymes PPO and POD are greater when the matrix studied contains *Stevia*.

These results show that *Stevia* could be useful for the food industry as a preservative agent and its combination with HHP treatment could be a good strategy to improve the microbial, nutritional and physico-chemical quality of foods.

1. INTRODUCTION

Browning of raw fruits, vegetables and beverages is a major problem for the food industry and it is believed to be one of the main causes of quality loss during postharvest handling and processing. Browning can cause deleterious changes in the appearance and organoleptic properties of foods, resulting in shorter shelf life, decreased market value and, sometimes, complete exclusion of the product from certain markets (Whitaker and Lee, 1995).

Polyphenoloxidase (PPO; monophenol, dihydroxy-L-phenylalanine oxygen oxidoreductase, EC 1.14.18.1) and peroxidase (POD; donor:hydrogen-peroxide oxidoreductase, EC 1.11.1.7) are responsible for the enzymatic browning reaction that occurs during the handling, storage and processing of fresh fruits and vegetables (Tomás-Barberán and Espín, 2001; Köksal and Gülçin, 2008). PPO converts phenolic compounds into dark coloured pigments, while POD catalyses the oxidation of a wide variety of organic and inorganic substrates in the presence of hydrogen peroxide generated in PPO-catalysed reactions. Therefore, the inactivation of these enzymes from harvest to consumer has a great impact in terms of nutritional value and acceptability.

A common approach for the prevention of browning of food and beverages has been, in addition to the physical methods to inactivate enzymes, the use of antibrowning agents that act primarily on the enzyme or react with the substrates and/or products of enzymatic catalysis in a manner that inhibits pigment formation (McEvily et al., 1992; Sapers and Miller, 1998; González-Aguilar et al., 2004; Zocca et al., 2011). In the presence of an antioxidant substance, such as ascorbic acid, or any other antibrowning agent

(4-hexylresorcinol, glutathione, L-cysteine, among others), PPO and POD activity might be depressed.

Food safety directs research in a constant quest for inhibitors from natural sources as alternatives to synthetic additives, because they are largely free of harmful side effects. Consequently, there is great demand from the food industry for safe, effective PPO inhibitors of natural origin. Recently, several studies have shown that natural agents have an inhibitory effect on PPO, including honey (Chen et al., 2000), onion extract (Lee et al., 2007), rice bran (Sukhonthara and Theerakulkait, 2012) and some aerial plant parts (Loizzo et al., 2012).

Consumers increasingly prefer food products that have a natural taste, flavour and colour and a nutritional value and that are safe. Nowadays it is well known that the use of preservatives is an effective way to ensure safety and stability of foods during storage time when combined with thermal and non-thermal processing techniques capable of reducing the microbial load present in raw materials. So far, thermal processing has been the main technology used by the food industry to achieve food preservation. In the past 40 years, however, alternative technologies that do not rely on heat have been developed, because intense heat produces undesirable changes in the nutritional and sensory properties of foods. These emerging non-thermal technologies normally allow food processing at room or refrigeration temperature (Corbo et al., 2009), and that is why they allow the production of safe, tasty, nutritious and fresh-like, minimally processed, ready-to-eat foods, so fashionable lately.

One of the outstanding technologies is high hydrostatic pressure (HHP) processing, adapted and adopted by food processors from the isostatic pressing process that has been used for a long time in the ceramics industry (San Martín et al., 2002). This is because of its ability to inactivate microorganisms and enzymes involved in food deterioration (Basak et al., 2002; Haiqiang and Hoover, 2003; Wang et al., 2012; Lae-Seung et al., 2013), and for this reason, in fact, many HHP-treated foods have already been introduced in the market (Butz and Tauscher, 2002), taking advantage of the fact that HHP processing is independent of the size and geometry of the product, which reduces the time required to process large quantities of food (Rastogi et al., 2007).

The beneficial effect of *Stevia* as an antioxidant, due to its composition, has been suggested by many researchers (Muanda et al., 2011), but to the best of our knowledge, there are no studies in the literature about the use of *Stevia* as an antibrowning agent, and therefore the potential use, practical implications and quantification of *Stevia* as an enzyme inhibitor is still a subject of research. Taking into account the *Stevia* properties mentioned, its combination with HHP treatment could be a good strategy to improve the microbial, nutritional and physico-chemical quality of foods.

Given that the antioxidant properties of *Stevia* have only been studied on the basis of chemical reactions (Halliwell et al., 1987; Germano et al., 2002; Kim et al., 2011), the antioxidant capacity of *Stevia* determined as PPO and POD percentage of inhibition was evaluated at 10 and 37°C. Moreover, the preservative potential of HHP processing and *Stevia* was investigated too, on the basis of their ability to inactivate *L. monocytogenes* and the oxidative enzymes PPO and POD. In both studies the matrix under study was a fruit extract containing orange, mango and papaya.

2. MATERIAL AND METHODS

2.1. Enzyme Crude Extract

Orange (*Citrus sinensis*, cv. Salustiana), mango (*Mangifera indica*) and papaya (*Carica papaya*) were purchased from a local supermarket and then stored at 4°C for one day before the enzyme extraction procedure.

POD and PPO were extracted using the method described by Rodrigo et al. (1996), with some modifications. Orange, mango and papaya fruits were washed, peeled and cut into small pieces. A mixture of the fruit pulp (orange, mango and papaya in a proportion of 15:20:65 (w/w/w), respectively) was homogenized in a proportion of 1:1 (w/v) with 0.05 M sodium phosphate buffer solution (Panreac Química, Spain) at pH 7.0 in a blender for 5 min at 4°C. The buffer contained 1 M NaCl (Scharlau Chemie, SA, Spain) and 5% (w/v) polyvinylpolypyrrolidone (Sigma-Aldrich®,USA). The homogenate was filtered through a layer of cheesecloth and the residue was centrifuged at 20,199 g for 30 min at 4°C with an Avanti J-25 centrifuge (Beckman Instruments, Inc., USA). The supernatant constituted the enzyme extract.

2.2. *Stevia* Infusion and Sample Preparation

Dried *Stevia* leaves were supplied by ANAGALIDE, SA (Spain) and stored at room temperature. To prepare a stock solution of 8.33 ± 0.01% (w/v), 100 mL of boiling mineral water was added to the dried leaves (8.33 g), and the mixture was covered and allowed to infuse for 30 min. The infusion was vacuum filtered using filter paper (Whatman® No. 1, Whatman International Ltd, UK) and the filtrate obtained was stored in 2 mL vials at –40 °C.

Different volumes of *Stevia* stock solution (1.2, 3.6 and 6 mL) were added to 14 mL of the crude enzyme extract to obtain final *Stevia* concentrations of 0.5, 1.5 and 2.5% (w/v), respectively. To evaluate the combined effect of *Stevia* with HHP the final *Stevia* concentrations used were 1.25 and 2.5% (w/v). The highest *Stevia* concentration (2.5% (w/v)) was selected, taking into account the sucrose concentration of commercial fruit-based beverages and the sweetness equivalence of *Stevia* and sucrose (Savita et al., 2004).

Water was added when necessary to give a final matrix volume of 20 mL. A blank sample was formulated with 14 mL of crude extract and 6 mL of water.

2.3. Determination of POD and PPO Activities

POD activity was measured spectrophotometrically, following the process reported by Rodrigo et al. (1996). Two hundred µL of extract was added to a tube with 3.8 mL of sodium phosphate buffer (0.05 M, pH 7.0 with 1 M NaCl), agitated for 5 min and kept at 25 °C for 7 min. The reaction began when 2.39 mL of this solution was placed in a polystyrene 4.5 mL cuvette and mixed with 0.61 mL of a reaction mixture consisting of 2-methoxyphenol (guaiacol; Sigma-Aldrich®, USA), H_2O_2 (Sigma-Aldrich®, USA) and bidistilled water.

PPO activity was measured by the method described by Giner et al. (2001), with some modifications. The method was based on measuring increase in absorbance at 410 nm when 1950 µL of 0.05 M 1,2-dihydroxybenzene (pyrocatechol; Sigma-Aldrich®, USA) in phosphate buffer (0.05 M, pH 7.00 with 1 M NaCl) as substrate reacted with 0.1 mL of enzyme extract.

Enzyme activity was calculated from the slope of the linear part of the graph of absorbance vs. time. Each enzyme determination was replicated three times at 25°C. The changes in absorbance, using a Lan Optics Model PG1800 spectrophotometer, (Labolan, Spain) were recorded every 1 s to 2.5 min with a Toshiba T1100 Plus computer (software UV-win® 5.0.5, PG Instruments Ltd, UK). One unit of POD or PPO activity was expressed as one absorbance increment (at 470 and 410 nm, respectively, in the conditions in which the assay was carried out) per min and mL of enzyme extract.

For both enzymes, the enzyme activity was expressed as relative activity (RA), calculated using the following equation:

$$RA = (A/A_0) \tag{1}$$

where A and A_0 are the current and initial POD and PPO activities, respectively.

In the case of HHP treatment, the enzyme activity was expressed as a percentage of relative activity (RA), calculated using the following equation:

$$\% RA = 100 \times (A/A_0) \tag{2}$$

where the enzyme activities of the fruit extract (A) were measured and compared with the initial activities of the untreated samples (A_0).

2.4. Microbial Culture Preparation

From a lyophilized pure culture provided by the Spanish Type Culture Collection, a stock vial containing *L. monocytogenes* (CECT 4032) was generated following the method described by Saucedo-Reyes et al. (2009). The average cell concentration was ca. 7.6×10^9 cfu/mL. It was established by viable plate count using TSA (Scharlau Chemie, SA, Spain) for the spreading of samples.

2.5. HHP Treatments

For each time and pressure condition, inoculated and un-inoculated samples of enzyme extract, with and without *Stevia*, were packed in polyethylene bags that were heat-sealed (MULTIVAC Thermosealer, Switzerland) before being inserted in the pressure vessel (High-Pressure Food Processor, EPSI NV, Belgium). HHP treatments were performed in a pilot-scale unit (2.35 L vessel volume), in a vessel with a maximum operating pressure of 600 MPa. The pressure transmitting fluid was a mixture of water and ethylene glycol (70:30% (v/v)). The samples were pressurized at 300, 400 and 500 MPa, at ambient temperature (18–22°C), for 5, 10 and 15 min. The pressure level, pressurization time and temperature were controlled automatically. The compression rate was 300 MPa/min and the decompression time was less than 1 min. The treatment time described in this study does not include come-up and come-down times. All the treatments were applied in triplicate. After completing the treatment, the samples were removed from the vessel and immediately transferred to an ice-water bath and stored under refrigeration ($3 \pm 1°C$) until needed for analysis.

POD and PPO activities and microbial inactivation of both untreated (blank) and HHP-treated samples were evaluated. Analyses were carried out immediately before and after the HHP treatments.

2.6. Microbial HHP Inactivation

The cellular density of treated samples, with and without *Stevia*, was determined, in terms of $\log_{10}$ (cfu/mL), before and after the different HHP treatments, by viable plate count, using TSA (Scharlau Chemie, SA, Spain) for sample spreading, followed by an incubation period of 48 h at 37°C.

The inactivation associated with each of the treatments tested was established according to the difference existing between the counts obtained pre- and post-treatment. From each sample, two aliquots were diluted and spread separately. The dilutions were done employing 1‰ (w/v) buffered peptone water (Scharlau Chemie, SA, Spain). From each aliquot, two plates were spread, so each count was obtained from four plates.

2.7. Statistical Analysis

Significant differences between the results were calculated by analysis of variance (ANOVA). Differences at $p < 0.05$ were considered to be significant.

3. RESULTS AND DISCUSSION

3.1. Effect of *Stevia* on Enzyme Activities in Fruit Extract

Before incubation, the POD and PPO activities of the extracts were 2.31 and 7.85 ΔAbsorbance/(min × mL of extract), respectively, in the absence of *Stevia*. These values were in the range of others described previously in the literature, although the POD and PPO contents of fruits and vegetables might be affected by species, cultivar and maturity (Castro et al., 2006; Elez-Martínez et al., 2006; Gasull and Becerra, 2006).

Enzyme activities in the fruit extract in the presence of different *Stevia* concentrations (0.5, 1.5 and 2.5% (w/v)) were 1.29 ± 0.18, 1.10 ± 0.25 and 1.45 ± 0.06 in relation to POD, and 6.91 ± 0.37, 6.77 ± 0.17 and 5.49 ± 0.13 with respect to PPO, respectively. According to these results, for both enzymes a significant decrease ($p < 0.05$) in fruit matrix enzyme activity was observed when *Stevia* was added. However, for POD, non-significant changes were obtained between the different *Stevia* concentrations, probably because a longer contact time between *Stevia* and the fruit extract would be needed for an effect to be observed.

3.2. Effect of *Stevia* on Enzyme Activities in Fruit Extract during Incubation

Samples with [0.5–2.5% (w/v)] and without *Stevia* were incubated in order to measure POD and PPO activities at different time intervals, depending on the incubation conditions and *Stevia* concentration, during 10 d or until the enzyme activity remained constant or below the detection limit. The incubation temperatures were 10°C, as a refrigerated temperature that depresses enzyme activity, and 37°C, as a near-optimal temperature for PPO and POD (Dincer et al., 2002; Mohamed et al., 2010).

Figure 1 shows the curves for POD and PPO degradation obtained by determining the residual activity of the samples studied. As expected, independently of *Stevia* concentration, the relative residual activity decreased with incubation time. For both enzymes the degradation level was always higher at 37 than at 10°C.

In the absence of *Stevia*, POD was more resistant to degradation than PPO.

This effect was on a larger scale at 37 °C. PPO activity was not detected after 2 d of incubation at 37°C, while slight POD activity (around 10%) was found after 6 d, at the same conditions. In relation to PPO, activity remained constant for 2 d at 10°C, but only for 4 hours at 37°C, decreasing to 60.4% of the initial activity during the 7th h of incubation, and it was not detected after the 2nd d of incubation. On the other hand, POD activity remained stable for 4 d at 10°C, whereas at 37°C it decreased to 69.7% in only 4 h and to 30%, from the initial value, after the 2nd d of incubation.

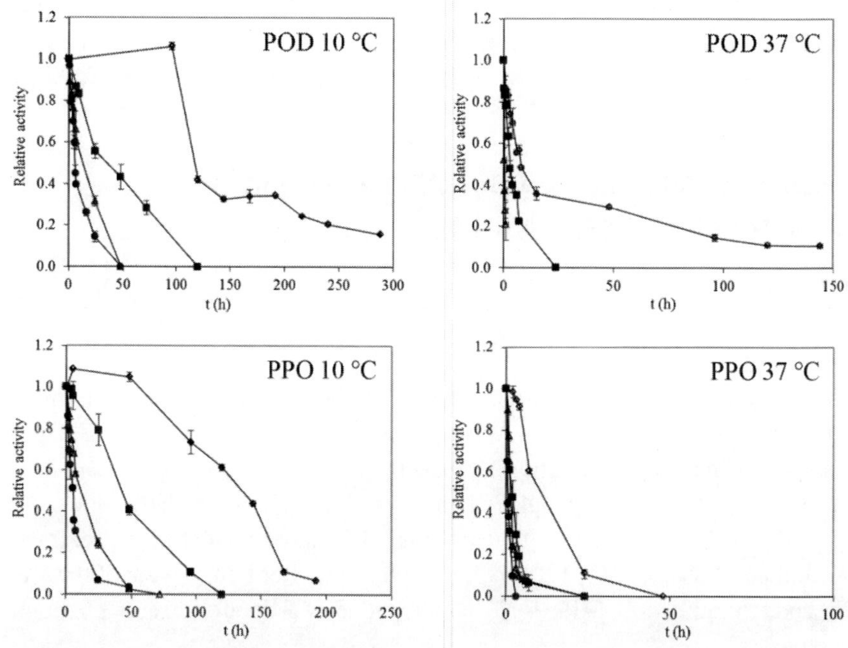

Figure 1. Effect of *Stevia* concentration (0% (w/v) (◊), 0.5% (w/v) (■), 1.5% (w/v) (Δ) and 2.5% (w/v) (●)) on POD and PPO activities at different temperatures (10 and 37°C).

The relative residual activity of POD and PPO decreased as the *Stevia* concentration increased. Moreover, as the *Stevia* concentration assayed increased, the activity reduction also increased. At 10 °C, PPO and POD activity decreased from 96% to 7% and from 99% to 14%, respectively, after 24 h of incubation, when 2.5% (w/v) of *Stevia* was added. At 37 °C, PPO activity was reduced from 98% to 9.5% after 2 hours of incubation, whereas POD was completely inhibited (activity below detection limit) when it was incubated with 2.5% (w/v) of *Stevia*.

This decrease in activity could be attributed to an inhibitory effect of *Stevia* on the oxidative enzymes POD and PPO. The mechanism of action for *Stevia* could be explained in relation to its constituents, which include ascorbic acid (14.98 mg/100 g of extract, dry basis) and also cinnamic acid and quercetin, which are some of the main phenolic and flavonol compounds present in *Stevia* leaves, respectively (Muanda et al., 2011; Wölwer-Rieck, 2012). It is known that PPO can be inhibited by ascorbic acid, because it reduces the quinone produced before it undergoes secondary reactions that lead to browning (Guerrero-Beltrán et al., 2005). In the case of undissociated forms of cinnamic acid and quercetin, they are able to inhibit PPO through complexation with copper at the active site of the enzyme (Kubo et al., 2000).

3.3. Effect of HHP and *Stevia* on PPO, POD and *L monocytogenes* in Fruit Extract

The effects of HHP treatments with and without *Stevia* on the PPO and POD activities and *L. monocytogenes* inactivation in the fruit extract are shown in Table 1.

With regard to oxidative enzymes, PPO and POD untreated extract activities in the absence of *Stevia* were 7.42±1.3 and 2.07±0.69 ΔAbsorbance/(min × mL of extract), respectively.

Pressure influenced residual PPO and POD activities. In the absence of *Stevia* the results showed that the residual PPO activity in the fruit extract decreased to 33.8%, 16.4%, 12.3%, 3.1% and 2.6% after treatment at 300 MPa for 5 and 15 min, 400 MPa for 10 min and 500 MPa for 5 and 15 min, respectively.

In the case of POD the residual activity values were 27.5%, 22.3%, 20.0%, 16.1% and 15.7%, at the same conditions, respectively, POD being more resistant to HHP treatments than PPO.

These differences were more pronounced in the 500 MPa treatments, where the residual POD activity in the fruit extract was 13% higher than the residual PPO.

Several studies that have focused on the effect of HHP on PPO and POD activities in various fruits and vegetables have shown that both enzymes exhibit a wide range of pressure stability. In general terms, the effectiveness of a treatment depends on the type of enzyme, pH, medium composition, temperature, time and pressure level applied (Hendrickx et al., 1998). However, many studies have demonstrated that PPO and POD present a rather high pressure resistance.

In relation to PPO from fruit smoothies, HHP treatments of 450 MPa for 5 min and 600 MPa for 10 min resulted in a 35% and 83% reduction in residual activity, respectively (Keenan et al., 2012).

With regard to POD, for green peas, a treatment of 900 MPa for 10 min at room temperature was needed to cause an 88% reduction in POD activity (Quaglia et al., 1996).

Table 1. Effect of HHP and *Stevia* on PPO, POD and *L. monocytogenes* in the fruit extract

Pressure (MPa)	Time (min)	*Stevia* (% (w/v))	PPO (% RA)	POD (% RA)	$\log_{10} (N_f/N_0)$ $\log_{10}$(cfu/mL)
300	5	0	33.84 ± 1.88	27.46 ± 2.69	0.17 ± 0.01
		2.5	15.60 ± 0.23	ND	3.96 ± 0.03
	10	1.25	17.96 ± 0.11	23.05 ± 1.24	-
	15	0	16.37 ± 0.23	22.29 ± 0.04	5.72 ± 0.02
		2.5	16.54 ± 0.69	ND	>7[+]
400	5	1.25	15.51 ± 0.11	20.55 ± 1.30	-
	10	0	12.25 ± 0.78	19.95 ± 0.73	6.30 ± 0.08
		1.25	12.35 ± 0.75[a]	18.52 ± 0.27[a]	-
		2.5	11.90 ± 0.36	ND	>7[+]
	15	1.25	12.07 ± 0.36	17.50 ± 0.01	-
500	5	0	3.14 ± 0.07	16.13 ± 0.59	>7[+]
		2.5	3.99 ± 0.05	ND	>7[+]
	10	1.25	3.00 ± 0.21	12.10 ± 0.01	-
	15	0	2.56 ± 0.21	15.66 ± 0.71	>7[+]
		2.5	1.87 ± 0.25	ND	>7[+]

[a]Average of central point. PPO: Polyphenoloxidase. POD: Peroxidase. ND: not detected. N_f: Final cell concentration. N_0: Initial cell concentration. -: Not tested. [+]$N_0 = 7.84 ± 0.17 \log_{10}$ (cfu/mL).

Owing to the very high pressure level that is often necessary to inactivate PPO and POD, several researches have proposed the addition of natural enzyme inhibitors, as the pressure inactivation of PPO and POD is influenced by the addition of salts, sugars or other compounds, including antibrowning agents. Table 1 shows, under the given experimental conditions, the capacity of *Stevia* to inhibit both PPO and POD enzymes.

To the best of our knowledge, this is the first time that the effect of *Stevia* on PPO and POD from fruits has been reported. The results obtained show that, in the case of POD, *Stevia* enhanced the inactivation percentage achieved by HHP. The three-way ANOVA showed that pressure and *Stevia* concentration had a significant influence on POD activity ($p < 0.05$). In the case of PPO, the residual activity present in extracts supplemented with 2.5% (w/v) and subjected to 300 MPa for 5 min (the lowest treatment assayed) was significantly less than the activity present in extract without *Stevia*, subjected to the same pressure and time combination ($p < 0.05$). Regardless of the HHP treatment applied, complete inactivation of POD was achieved when the *Stevia* concentration was 2.5% (w/v). In the case of PPO, the maximum reduction achieved was 98.13% with the highest treatment tested (500 MPa, 15 min, 2.5% (w/v) *Stevia*). In this case, since the enzyme-antibrowning agent interaction might alter the pressure stability of the enzyme, the combination of chemical and physical browning prevention treatments would enable the use of less intense inactivation treatments.

In view of the results obtained (Table 1), it can be concluded that HHP is an appropriate tool to reduce the concentration of *L. monocytogenes* in the matrix developed. For a given pressure, the microbial inactivation was higher when the treatment time was increased, independently of the *Stevia* concentration used in the sample formulation. In addition, pressures higher than 300 MPa or times higher than 5 min were always able to inactivate at least 5 log cycles, the standard proposed for any processing strategy intended to guarantee the safety of fruit juices and similar products (CFR, 2012), *L. monocytogenes* being the pertinent pathogen.

For any pressure-time combination, in the presence of *Stevia* the number of inactivated log cycles was always greater than those achieved in its absence (Table 1). For instance, the addition of 2.5% (w/v) *Stevia* increased the inactivation reached after 300 MPa, 5 min, by almost 4 log cycles. This is due to the antimicrobial nature of *Stevia*, which has also been reported by other authors (Sivaram and Mukundan, 2003; Tadhani and Subhash, 2006; Debnath, 2008; Ghosh et al., 2008; Jayaraman et al., 2008; Seema, 2010).

The results suggested that the combination of HHP with *Stevia* can be a useful tool to obtain safe and shelf-stable fruit-based beverages. Specifically, a treatment of 500 MPa applied for 15 min combined with 2.5% (w/v) *Stevia* inactivates *L. monocytogenes* by more than 5 log cycles while it reduces PPO activity to 1.87% and POD below the detection limit.

CONCLUSION

According to the results obtained, *Stevia* has shown an antioxidant activity based on the inhibition of oxidative enzymes, so it could potentially be used as a PPO and POD natural antibrowning agent to improve the preservation of fresh-cut vegetables and fruits.

The combined use of HHP with *Stevia* showed that the application can be a useful tool in order to obtain safe and shelf-stable fruit-based beverages with enhanced antioxidant and nutritional properties, given that *Stevia* is a natural ingredient that could act both as an antioxidant and as an antimicrobial. Therefore, the current use of this natural product to sweeten foodstuffs and beverages could be reassessed in the light of the additional preservative properties shown if a cold chain failure occurs or if foods are kept incorrectly, especially against pathogens such as *L. monocytogenes*, or against enzymes such as PPO and POD, which are responsible for the main oxidative reactions.

ACKNOWLEDGMENTS

The authors thank the Ministry of Economy and Competitiveness for its support through project number AGL2010-22206-C02-01-ALI and the company ANAGALIDE, SA (Spain) for providing them with dried *Stevia* leaves. María Nieves Criado and Clara Miracle Belda-Galbis are especially grateful to the CSIC for providing them with a JAE-Tec and a JAE-Predoc grant, respectively.

REFERENCES

Basak, S., Ramaswamy, H. S. & Piette, J. P. G. (2002). High pressure destruction kinetics of *Leuconostoc mesenteroides* and *Saccharomyces cerevisiae* in single strength and concentrated orange juice. *Innovative Food Science & Emerging Technologies, 3,* 223-231.

Butz, P. & Tauscher, B. (2002). Emerging technologies: Chemical aspects. *Food Research International, 35,* 279-284.

Castro, J. A., Baquero, L. E. & Narváez, C. E. (2006). Catalase, peroxidase and polyphenoloxidase from pitahaya amarilla fruits (*Acanthocereus pitajaya*). *Revista Colombiana de Química, 35,* 91-100.

CFR (2012). Title 21: Food and drugs. USA:FDA. Available in: http://www.accessdata.fda.gov/scripts/cdrh/cfdocs/cfcfr/CFRSearch.cfm?fr=120.24.

Chen, L., Mehta, A., Berenbaum, M., Zangerl, A. R. & Engeseth, N. J. (2000). Honeys from different floral sources as inhibitors of enzymatic browning in fruit and vegetable homogenates. *Journal of Agricultural and Food Chemistry, 48,* 4997-5000.

Corbo, M. R., Bevilacqua, A., Campaniello, D., D'Amato, D., Speranza, B. & Sinigaglia, M. (2009). Prolonging microbial shelf life of foods through the use of natural compounds and non-thermal approaches - A review. *International Journal of Food Science & Technology, 44,* 223-241.

Debnath, M. (2008). Clonal propagation and antimicrobial activity of an endemic medicinal plant *Stevia rebaudiana. Journal of Medicinal Plants Research, 2,* 45-51.

Dincer, B., Colak, A., Aydin, N., Kadioglu, A. & Guner, S. (2002). Characterization of polyphenoloxidase from medlar fruits (*Mespilus germanica* L., Rosaceae). *Food Chemistry, 77,* 1-7.

Elez-Martínez, P., Soliva-Fortuny, R. C. & Martín-Belloso, O. (2006). Comparative study on shelf life of orange juice processed by high intensity pulsed electric fields or heat treatment. *European Food Research and Technology, 222,* 321-329.

Gasull, E. & Becerra, D. (2006). Characterization of polyphenoloxidase extracted from pears (cv. Packam′s Triumph) and apples (cv. Red Delicious). *Información tecnológica, 17,* 69-74.

Germano, M. P., De Pasquale, R., D'Angelo, V., Catania, S., Silvari, V. & Costa, C. (2002). Evaluation of extracts and isolated fraction from *Capparis spinosa* L. buds as an antioxidant source. *Journal of Agricultural and Food Chemistry, 50,* 1168-1171.

Ghosh, S., Subudhi, E. & Nayak, S. (2008). Antimicrobial assay of *Stevia rebaudiana* Bertoni leaf extracts against 10 pathogens. *International Journal of Integrative Biology, 2,* 27-31.

Giner, J., Gimeno, V., Barbosa-Cánovas, G. V. & Martín, O. (2001). Effects of pulsed electric field processing on apple and pear polyphenoloxidases. *Food Science and Technology International, 7,* 339-345.

González-Aguilar, G. A., Ruiz-Cruz, S., Cruz-Valenzuela, R., Rodríguez-Félix, A. & Wang, C. Y. (2004). Physiological and quality changes of fresh-cut pineapple treated with antibrowning agents. *LWT - Food Science and Technology, 37,* 369-376.

Guerrero-Beltrán, J. A., Swanson, B. G. & Barbosa-Cánovas, G. V. (2005). Inhibition of polyphenoloxidase in mango puree with 4-hexylresorcinol, cysteine and ascorbic acid. *LWT - Food Science and Technology, 38,* 625-630.

Haiqiang, C. & Hoover, D. G. (2003). Modeling the combined effect of high hydrostatic pressure and mild heat on the inactivation kinetics of *Listeria monocytogenes* Scott A in whole milk. *Innovative Food Science & Emerging Technologies, 4,* 25-34.

Halliwell, B., Gutteridge, J. M. C. & Aruoma, O. I. (1987). The deoxyribose method: A simple "test-tube" assay for determination of rate constants for reactions of hydroxyl radicals. *Analytical Biochemistry, 165,* 215-219.

Hendrickx, M., Ludikhuyze, L., Van den Broeck, I. & Weemaes, C. (1998). Effects of high pressure on enzymes related to food quality. *Trends in Food Science & Technology, 9,* 197-203.

Jayaraman, S., Manoharan, M. S. & Illanchezian, S. (2008). *In-vitro* antimicrobial and antitumor activities of *Stevia rebaudiana* (Asteraceae) leaf extracts. *Tropical Journal of Pharmaceutical Research, 7,* 1143-1149.

Keenan, D. F., Roessle, C., Gormley, R., Butler, F. & Brunton, N. P. (2012). Effect of high hydrostatic pressure and thermal processing on the nutritional quality and enzyme activity of fruit smoothies. *LWT - Food Science and Technology, 45,* 50-57.

Kim, I., Yang, M., Lee, O. & Kang, S. (2011). The antioxidant activity and the bioactive compound content of *Stevia rebaudiana* water extracts. *LWT - Food Science and Technology, 44,* 1328-1332.

Köksal, E. & Gülçin, I. (2008). Purification and characterization of peroxidase from cauliflower (*Brassica oleracea* L. var. botrytis) buds. *Protein and Peptide Letters, 15,* 320-326.

Kubo, I., Kinst-Hori, I., Chaudhuri, S. K., Kubo, Y., Sanchez, Y. & Ogura, T. (2000). Flavonols from Heterotheca inuloides: Tyrosinase inhibitory activity and structural criteria. *Bioorganic & Medicinal Chemistry, 8,* 1749-1755.

Lae-Seung, J., Seung Hwan, L., Sungkyun, K. & Juhee, A. (2013). Effect of high hydrostatic pressure on the quality-related properties of carrot and spinach. *Food Science and Biotechnology, 22,* 189-195.

Lee, M. Y., Lee, M. K. & Park, I. (2007). Inhibitory effect of onion extract on polyphenol oxidase and enzymatic browning of taro (*Colocasia antiquorum* var. esculenta). *Food Chemistry, 105,* 528-532.

Loizzo, M. R., Tundis, R. & Menichini, F. (2012). Natural and synthetic tyrosinase inhibitors as antibrowning agents: An update. *Comprehensive Reviews in Food Science and Food Safety, 11,* 378-398.

McEvily, A. J., Iyengar, R. & Otwell, W. S. (1992). Inhibition of enzymatic browning in foods and beverages. *Critical Reviews in Food Science and Nutrition, 32,* 253-273.

Mohamed, S. A., El-Badry, M. O., Drees, E. A. & Fahmy, A. S. (2010). Properties of a cationic peroxidase from *Citrus jambhiri* cv. Adalia. *Applied Biochemistry and Biotechnology, 162,* 926-926.

Muanda, F. N., Soulimani, R., Diop, B. & Dicko, A. (2011). Study on chemical composition and biological activities of essential oil and extracts from *Stevia rebaudiana* Bertoni leaves. *LWT - Food Science and Technology, 44,* 1865-1872.

Quaglia, G. B., Gravina, R., Paperi, R. & Paoletti, F. (1996). Effect of high pressure treatments on peroxidase activity, ascorbic acid content and texture in green peas. *LWT - Food Science and Technology, 29,* 552-555.

Rastogi, N. K., Raghavarao, K. S. M. S., Balasubramaniam, V. M., Niranjan, K. & Knorr, D. (2007). Opportunities and challenges in high pressure processing of foods. *Critical Reviews in Food Science and Nutrition, 47,* 69-112.

Rodrigo, C., Rodrigo, M., Alvarruiz, A. & Frígola, A. (1996). Thermal inactivation at high temperatures and regeneration of green asparagus peroxidase. *Journal of Food Protection, 59,* 1065-1071.

San Martín, M. F., Barbosa-Cánovas, G. V. & Swanson, B. G. (2002). Food processing by high hydrostatic pressure. *Critical Reviews in Food Science and Nutrition, 42,* 627-645.

Sapers, G. M. & Miller, R. L. (1998). Browning inhibition in fresh-cut pears. *Journal of Food Science, 63,* 342-346.

Saucedo-Reyes, D., Marco-Celdrán, A., Pina-Pérez, M. C., Rodrigo, D. & Martínez-López, A. (2009). Modeling survival of high hydrostatic pressure treated stationary- and exponential-phase *Listeria innocua* cells. *Innovative Food Science & Emerging Technologies, 10,* 135-141.

Savita, S. M., Sheela, K., Sunanda, S., Shankar, A. G., Ramakrishna, P. & Sakey, S. (2004). Health implications of *Stevia rebaudiana*. *Journal of Human Ecology, 15,* 191-194.

Seema, T. (2010). *Stevia rebaudiana*: A medicinal and nutraceutical plant and sweet gold for diabetic patients. *International Journal of Pharmacy & Life Sciences, 1,* 451-457.

Sivaram, L. & Mukundan, U. (2003). *In vitro* culture studies on *Stevia rebaudiana*. *In vitro Cellular and Developmental Biology-Plant, 39,* 520-523.

Sukhonthara, S. & Theerakulkait, C. (2012). Inhibitory effect of rice bran extract on polyphenol oxidase of potato and banana. *International Journal of Food Science and Technology, 47,* 482-487.

Tadhani, M. B. & Subhash, R. (2006). *In vitro* antimicrobial activity of *Stevia rebaudiana* Bertoni leaves. *Tropical Journal of Pharmaceutical Research, 5,* 557-560.

Tomás-Barberán, F. & Espín, J. C. (2001). Phenolic compounds and related enzymes as determinants of quality in fruits and vegetables. *Journal of the Science of Food and Agriculture, 81,* 853-876.

Wang, R., Wang, T., Zheng, Q., Hu, X., Zhang, Y. & Liao, X. (2012). Effects of high hydrostatic pressure on color of spinach purée and related properties. *Journal of the Science of Food and Agriculture, 92,* 1417-1423.

Whitaker, J. R. & Lee, C. Y. (1995). In Whitaker, J. R. & Lee, C. Y. (Eds.), *Enzymatic browning* (pp. 2-7). USA:American Chemical Society Symposium.

Wölwer-Rieck, U. (2012). The leaves of *Stevia rebaudiana* (Bertoni), their constituents and the analyses thereof: A Review. *Journal of Agricultural and Food Chemistry, 60,* 886-895.

Zocca, F., Lomolino, G. & Lante, A. (2011). Dog rose and pomegranate extracts as agents to control enzymatic browning. *Food Research International, 44,* 957-963.

In: Stevia rebaudiana
Editors: D. Betancur and M. Segura

ISBN: 978-1-63463-335-2
© 2015 Nova Science Publishers, Inc.

Chapter 6

CHEMICAL COMPOSITION AND ANTIOXIDANT CAPACITY OF *STEVIA REBAUDIANA* (VAR. MORITA II) LEAVES FROM YUCATAN, MÉXICO

Y. B. Moguel-Ordóñez[1], D. L. Cabrera-Amaro[1], D. A. Betancur-Ancona[2], M. R. Segura-Campos[2] and J. C. Ruiz-Ruiz[3,]*

[1]Instituto Nacional de Investigaciones Forestales, Agrícolas y Pecuarias. Mocochá, Yucatán, México
[2]Facultad de Ingeniería Química, Universidad Autónoma de Yucatán, Chuburná de Hidalgo Inn, Mérida, Yucatán, México
[3]Departamento de Ingeniería Química-Bioquímica, Instituto Tecnológico de Mérida, Mérida, Yucatán, México

ABSTRACT

This study was carried out to evaluate the primary nutrients and dietetic fiber composition, as well as to determine the functional properties, of the *Stevia rebaudiana* (Bertoni) leaves when dried using one of two methods (convective drying and shade drying). Between the drying treatments, no differences were found for moisture content (7.5–

---

* jcruiz_ruiz@hotmail.com.

7.7%), ash content (7.7–7.9%), crude fat (3.2–3.8%) or crude fiber (10.5–10.0%); however, crude protein (12.1–14.6%) and nitrogen-free extract (59.0–56.0%) showed significant differences. With regard to fiber fractions, these showed differences among neutral detergent fiber (17.8–16.2%), acid detergent fiber (13.7–12.3%) and cellulose (11.4–9.8%); on the other hand, acid detergent lignin (2.3–2.5%) and hemicellulose (4.1–3.9%) did not show differences. The results of total dietary fiber (38.2–32.8%) were equal for both drying treatments; however, there were differences in soluble dietary fiber (6.4–3.7%) and in insoluble dietary fiber (31.7–29.1%). Chlorophyll a and b contents showed significant differences between drying treatments, with values of 0.030–0.034% and 0.034–0.039% for chlorophyll a and b respectively. Carotenoid content did not show significant differences between drying treatments: 0.065% (convection dried) and 0.06% (shade dried). Both kinds of pigments are related to the antioxidant capacity of the extracts; in this sense, the Trolox equivalent antioxidant capacity values (417–423 mM/mg sample), ferric reducing power values (85–88%), copper chelating capacity (59–60%), and ferric chelating capacity (49–52%) did not show significant differences between drying treatments. The results indicate that stevia leaves dried using either method may be an ideal candidate for further research into their uses as a source of nutrients, fiber and antioxidant compounds.

INTRODUCTION

The genus Stevia belongs to the Asteraceae family, tribe Eupatoriae, and comprises 240 species, growing mostly at the altitude of 500–3000 m in semi-dry mountainous terrain. Various species of Stevia contain several potential sweetening compounds, with *Stevia rebaudiana* (Bertoni) being the sweetest of all (Singh et al., 2012). Stevia (*Stevia rebaudiana*, Bertoni) is a perennial shrub that is indigenous to Paraguay and Brazil. Nowadays, the extraction of sweeteners from stevia leaves is a growing industrial and commercial sector worldwide; more than 750 tons of stevia leaves per year are used as a crude extract for consumption and extraction of glycosides (Atteh et al., 2011). The sweetening property is associated with their contents of several glycosides: stevioside, steviobioside, rebaudiosides A to F, dulcoside A and steviol. These glycosides and their derivatives are known to account for 4–20% of the dry weight of stevia leaves (Geuns et al., 2003). Stevia and its extract have been studied widely from the sweetener point of view; however, a search through literature shows no information on the non-sweetening components that make up 80–90% dry weight of this plant. The non-sweet constituents identified in

S. rebaudiana leaves are: diterpenes, triterpenes, sterols, flavonoids, volatile oil constituents, pigments and inorganic matters (Shukla et al., 2009). Some of these components are phytochemicals that could act as antioxidants.

The role of reactive oxygen species (ROS) has been determined in many human degenerative diseases, including ageing, cancer, arthritis and Parkinson's disease (Carreras et al 2004). For example, hydrogen peroxide (H_2O_2), a prominent ROS, causes lipid peroxidation and DNA damage in cells (Cho et al., 2006). The antioxidant action of some natural compounds, such as vitamins and minerals, polyphenols and other non-nutrient compounds of plants, inhibiting the generation of reactive oxygen species or the scavenging of free radicals, is believed to be beneficial for human health. Indeed, natural antioxidants have displayed a wide range of pharmacological activity, such as anti-cancer, anti-inflammatory and anti-ageing actions (Pinnell, 2003). To counteract this threat to their integrity, cells have evolved a variety of defense systems based on both water-soluble and lipid-soluble antioxidant species, and on antioxidant enzymes. A high proportion of the antioxidant systems of the human body are dependent on dietary constituents (Nehir and Karakaya, 2004). Synthetic antioxidants such as butylhydroxyanisole (BHA) or butylhydroxytoluene (BHT) are used to decelerate these processes. However, due to their unstable and highly volatile nature, they have frequently raised questions about their safety and efficiency, ever since their introduction into the food industry (Ebrahimzadeh et al., 2010). Consequently, the need to identify alternative natural and safe sources of antioxidants arose, and the search for safe and natural antioxidants, especially of plant origin, has notably increased in recent years (Mahmoudi et al., 2009).

The steviol glycosides extracted and purified from the leaves of *S. rebaudiana* are the most sold product of this plant. However, whole leaves dried, crushed leaves, leaves for tea and for industrial use are also sold. Besides steviol glycosides, stevia leaves also contain amino acids, carbohydrates, enzymes, organic acids, polysaccharides, vitamins A, B2 and B6, carotene and microelements, among other components like phytochemicals (Atteh et al., 2011). These compounds may be important for the nutrition and health of people who consume the products obtained from this plant. The drying of the *S. rebaudiana* leaves is important in order to preserve them during storage and transportation, but the method used influences the concentration of chemical constituents and must be done within 24 hours after harvesting (Abou-Arab et al., 2010). The methods most used for drying stevia leaves include the use of equipment or traditional drying, carried out under ambient conditions Espitia et al. (2009). Conventional drying with hot air is a commonly-used process of dehydratation. However, an alternative way to remove excess water from food is by using traditional methods like shade drying and sun drying, which are drying processes based on water evaporation under natural environmental conditions (Verás et al., 2012). The success of drying as a method of food preservation depends on the properties of the dried products, such as color, nutritive value and biological activities (Bhuiyan et al., 2011).

S. rebaudiana leaves can be a potential source of nutrients and natural antioxidants. The bioactivity of any plant products varies greatly with the change of geographical conditions, such as soil, water cultivation process, post-harvest treatments, etcetera. The aim of this study was to evaluate the influence of drying treatments on the primary nutrients, dietetic fiber composition and antioxidant capacities of *Stevia rebaudiana* (Bertoni) leaves.

MATERIALS AND METHODS

Stevia rebaudiana Bertoni (var. Morita II) was obtained from plots established in Tizimín, Yucatán, México. The plantation had a crop managed in accordance with the production technology described by Ramirez et al., (2011). Samples were obtained from the first cut of the plot at three months of age; samples consisted of branches of leaves and stems. All chemicals were reagent grade or better, and were purchased from Sigma Chemical Co. (St. Louis, MO, USA).

Drying Treatments

Stevia rebaudiana leaves were subjected to two treatments of drying: convective drying and shade drying. An oven with a temperature setting of 60 °C was used for the convective drying treatment. For the shade drying treatment, temperature and relative humidity were monitored during the entire process. The vegetal material was distributed in aluminum trays, with three replicates for each treatment. To establish the kinetics of drying, the weight of the leaves was recorded every 4 h (during 24 h) for the convective drying treatment and every 24 h (during 96 h) for the shade drying treatment. Drying was stopped until the leaves reached a stable moisture content.

Proximate Composition

The proximate composition of *S. rebaudiana* leaves was determined using AOAC (1998) methods: moisture content (Method 925.09), ash (Method 923.03), crude fat (Method 920.39), crude protein using a 6.25 nitrogen-protein conversion factor (Method 954.01), and crude fiber (Method 962.09). The carbohydrate content was determined by subtracting the total crude protein, crude fiber, ash and crude fat from the total dry weight (100 g), and was estimated as the nitrogen-free extract (NFE).

Total Dietary Fiber (TDF)

This parameter was determined with the gravimetric enzymatic method proposed by Prosky et al. (1988). Briefly, 1 g of sample was weighed into each of four flasks, and 50 mL of phosphate buffer (50.0 mM, pH 6) were added to each. The flasks were then placed in a water bath at 100 °C, 0.1 mL of thermostable α-amylase enzyme (Sigma A–3306) was added to each, and then they were agitated at 60 rpm for 15 min. After cooling, pH was adjusted to 7.5. The flasks were returned to the bath at 60 °C, 0.1 mL protease (Sigma P–3910) was added to each, and then they were agitated at 60 rpm for 30 min.

After cooling, pH was adjusted to 4.0. The flasks were again placed in the bath at 60 °C, 0.3 mL amyloglucosidase (Sigma A–9913) was added to each, and then they were agitated at 100 rpm for 30 min. Finally, ethanol 95% (v/v), preheated to 60 °C, was added at a 1:4 (v/v) ratio. In a vacuum, flask content was filtered into crucibles containing celite (Sigma C–8656). The residue remaining in the flask was washed three times with 20 mL of ethanol 78% (v/v), twice with 10 mL of ethanol 95% (v/v) and twice with 10 mL of acetone.

Crucible content was dried at 105 °C. Protein (N x 6.25) was determined for the residue in two crucibles, and the residue in the remaining two was burned at 550 °C for 4 h.

$$TDF\ (\%) = \frac{[Residue\ weight\ (g) - protein\ (g) - ash\ (g)] \times 1000}{Sample\ weight\ (g)}$$

Insoluble Dietary Fiber (IDF)

This was determined following the method of Prosky et al. (1988), which is similar to that for TDF, except that addition of ethanol 95% (v/v) at 1:4 (v/v) is omitted.

Soluble Dietary Fiber (SDF)

Calculated by the difference between TDF and IDF:

$$SDF = TDF - IDF.$$

Fiber Components

The fiber components were determined using the methods of Van Soest et al. (1991):

Neutral Detergent Fiber (NDF)

50 mL glass crucibles were dried at 100 °C overnight and their weights were recorded (W1). Approximately 1 g samples were weighed (W2) into a Berzelius baker. A volume of 100 mL neutral-detergent solution was added at room temperature. The solution was prepared with distilled water, sodium borate, EDTA, lauryl sulfate, 2-ethoxyethanol and disodium phosphate. The pH was adjusted at 6.9–7.1. The samples were heated to boiling in 5–10 min; heat was reduced to avoid foaming as boiling commenced. The samples were refluxed 60 min from onset of boil. The samples were filtered through the glass crucibles using vacuum, and were washed with boiling water three times. The residue was washed twice with acetone (40 mL). Finally, the residue was dried overnight in a convection oven at 100 ˚C. The weight was then recorded (W3). The percentage of Neutral Detergent Fiber (NDF) was calculated as follows: %NDF = (W3–W1/W2) x 100
where:

W1 = tare weight of crucible in grams

W2 = initial sample weight in grams

W3 = dry weight of crucible and dry fiber in grams.

Acid Detergent Fiber (ADF)

Samples were processed in a similar way to the method reported for NDF, using a volume of 100 mL of acid detergent solution. The solution was prepared with distilled water and 20 g of cetyl-trimethylammonium bromide in 1 L of sulfuric acid 1.0 N.

Acid Detergent Lignin (ADL)

Samples were processed in a similar way to the method reported for ADF, using a volume of 100 mL 72% H_2SO_4 solution. The crucible with sample was burned in a muffle furnace at 500°C for 2 hours. The crucible was weighed to the nearest 0.1 mg (W3).

Cellulose and Hemicellulose

These parameters were calculated as follows:

%Cellulose = ADF (Cellulose, lignin, cutin) − ADL (Lignin, cutin)

%Hemicellulose = NDF (Hemicellulose, cellulose, lignin, cutin) − ADF (Cellulose, lignin, cutin).

Determination of Chlorophylls (a and b) and Carotenoids Contents

The weighed samples, having been put separately in 90% acetone (1.0 mL for each 50 mg), were homogenized at 1000 rpm for 1 min. The homogenate was filtered and centrifuged at 2500 rpm for 10 min. The supernatant was separated and the absorbances were read at 665, 645, 630 and 444 nm, respectively. The amount of chlorophylls and carotenoids were calculated according to the formulas of Lichtenthaller (1987):

Chlorophyll a (%) = 11.6*A665 - 1.31*A645 – 0.14*A630

Chlorophyll b (%) = –4.34*A664 + 20.7*A645 – 4.42*A630

Total pigments (%) = Chlorophyll a + Chlorophyll b + Carotenoids

In Vitro Antioxidant Properties of *Stevia Rebaudiana* Extracts 2,2´-Azinobis-(3-Ethylbenzothiazoline-6-Sulfonic Acid, ABTS +) Decolorization Assay

The ABTS•+ radical cation was produced by reacting 2,2'-azino-bis(3-ethylbenzothiazoline-6-sulphonic acid (ABTS) with potassium persulfate (Rhee et al., 2004). To prepare the stock solution, ABTS was dissolved at a 2 mM concentration in 50 mL of phosphate-buffered saline. The ABTS radical cation was produced by reacting 10 mL ABTS stock solution with 40 μL $K_2S_4O_8$ 70 mM solution, and allowing the mixture to stand in darkness at room temperature for 16–17 h before use. The radical was stable in this form for more than 2 days when stored in darkness at room temperature. Antioxidant compound content in the stevia stems was analyzed by diluting the ABTS•+ solution with PBS to an absorbance of 0.800 ± 0.030 AU at 734 nm.

After adding 990 μL diluted ABTS•+ solution (A 734 nm = 0.800 ± 0.030) to 10 μL of 6-hydroxy-2,5,7,8-tetramethylchroman-2-carboxylic acid (TROLOX) standard (final concentration 0.5-3.5 mM) in PBS, absorbance was read at room temperature exactly 6 min after initial mixing. All analyses were run in triplicate. The percentage decrease in absorbance at 734 nm was calculated and plotted as a function of the antioxidant concentration, or of Trolox for the standard reference data. To calculate the Trolox equivalent antioxidant coefficient (TEAC), the slope of the absorbance inhibition percentage vs. antioxidant concentration plot was divided by the slope of the Trolox plot. This produced the TEAC at a specific point in time.

Ferric Reducing Power Assay

This method is based on the reduction of potassium ferricyanide (Fe^{+3}) to (F^{+2}) in the presence of an antioxidant, KFeIII forming the blue complex [FeII(CN$_6$)], which absorbed at 700 nm (Sudha et al., 2011). First, 200 µL of sample (containing 1 mg of protein), 500 µL of phosphate buffer (0.2 M, pH 6.6), and 500 µL of potassium ferricyanide (1%) were mixed in a test tube. The test tube was then incubated at 50 °C for 20 min. Subsequently, 500 mL of trichloroacetic acid (10%) were added, and the tube was centrifuged at 3000 g for 10 min. An aliquot of 500 µL of the supernatant was dissolved in an equal amount of distilled water, and immediately 500 µL of ferric chloride (0.1%) were added. Absorbance was determined at 700 nm. Samples were tested in a range of concentrations from 200 to 1000 mg/mL. Butylated hydroxytoluene was used as control in the same range of concentrations.

Chelating Of Metal Ions Cu^{2+} and Fe^{2+}

Cu^{2+}-chelating activity was determined using the pyrocatechol violet reagent in accordance with Saiga et al. (2003). Briefly, 1.0 mL of sodium acetate buffer (100 mM, pH 4.9), 100 mL of Cu (II) standard solution (1.0 mg/mL), and 100 mL of sample (containing 200 mg of sample) were mixed in a test tube. The mixture was allowed to react for 5 min at room temperature, after which 25 mL of a pyrocatechol violet solution (4.0 mM) was added. Absorbance was determined at 632 nm. Copper chelating activity was calculated as follows:

$$\text{Chelating activity (\%)} = (1 - \text{sample absorbance/control absorbance}) \times 100.$$

Fe^{2+}-chelating activity was determined by measuring the formation of the Fe^{2+}-ferrozine complex according to Carter (1971). Briefly, 1.0 mL of sodium acetate buffer (100 mM, pH 4.9), 100 mL of Fe(II) standard solution (1.0 mg/mL), and 100 mL of sample (containing 200 mg of sample) were mixed in a test tube. The mixture was allowed to react for 5 min at room temperature, after which 50 mL of a ferrozine solution (40 mM) was added. Absorbance was determined at 562 nm. Iron chelating activity was calculated as follows:

Chelating activity (%) = (1 − sample absorbance/ control absorbance) x 100.

STATISTICAL ANALYSIS

All results were analyzed using descriptive statistics with a central tendency and dispersion measures. One-way ANOVAs were run to evaluate proximate composition, fiber composition, pigments content and *in vitro* antioxidant activities. The least significant difference (LSD) multiple range test was used to determine differences among drying treatments. All analyses were conducted in accordance with Montgomery (2004), and processed with the Statgraphics Plus version 5.1 software.

RESULTS AND DISCUSSION

Drying Treatments

Drying curves obtained with convective treatments are shown in Figures 1 and 2. For convective drying, a gradual decrease of the weight difference was observed during the first 20 hours, after which the weight difference stabilized because the vegetal material reached equilibrium in the moisture content. Dehydration of leaves was also evaluated using the traditional method of shade drying. For this method, a rapid decrease of the weight difference was observed during the first 20 hours, after which the weight difference stabilized because the vegetal material reached equilibrium in the moisture content.

Both treatments were equally effective in drying the leaves. Shade drying treatment depends on the weather conditions to dehydrate the leaves, which reduces costs since this method does not require equipment or energy. However, the dependence of this treatment on weather conditions does not allow for the establishing of reproducible conditions to control the drying process. Drying is important in order to maintain the leaves during storage and transport, and also has an influence on the concentration of the nutritive and functional components.

Proximate Analysis

Table 1 summarizes the proximate composition of *S. rebaudiana* leaves dehydrated using two different drying methods. Both methods reduced the moisture content to values less than 10%. The moisture content varied from 7.45% for convection-dried leaves to 7.72% for shade-dried leaves.

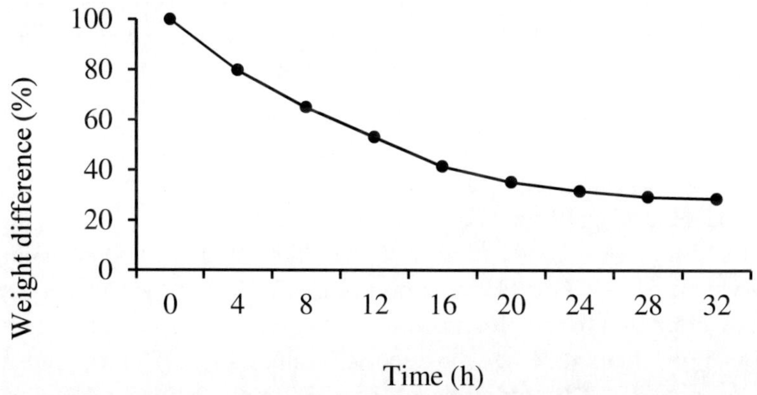

Figure 1. Difference of weight (db) in leaves of Stevia rebaudiana Bertoni obtained by convective drying treatment. Data are presented as means (n = 3).

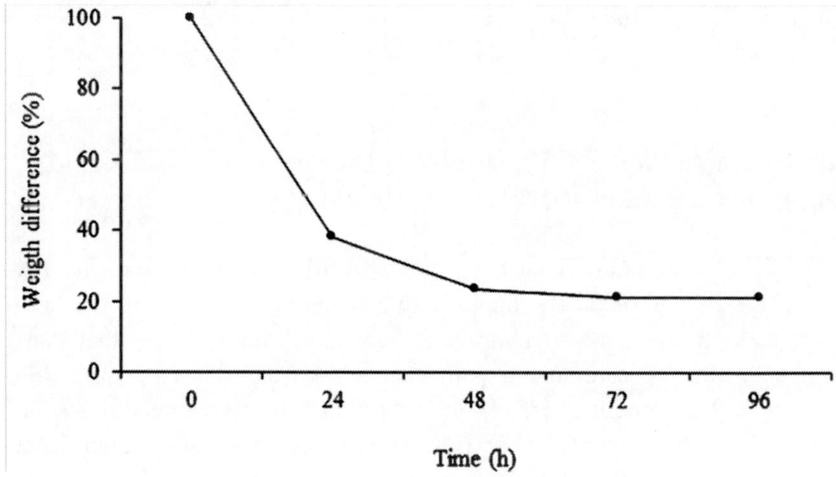

Figure 2. Difference of weight (db) in leaves of Stevia rebaudiana Bertoni obtained by shade drying treatment. Data are presented as means (n = 3).

Table 1. Proximate composition of *Stevia rebaudiana* (Bertoni) var. Morita II leaves treated by convective and shade drying

Component (%)	Convection dried	Shade dried
Moisture	7.45 ± 0.10^a	7.72 ± 0.24^a
Ash	7.73 ± 0.29^a	7.89 ± 0.28^a
Crude Fat	3.23 ± 0.41^a	3.81 ± 0.64^a
Crude Protein	12.11 ± 0.66^a	14.57 ± 0.38^b
Crude fiber	10.50 ± 0.36^a	10.00 ± 0.14^a
NFE	66.43 ± 1.36^a	63.73 ± 0.79^b

[a] Data are presented as means (n = 3).
[b] Different letters in the same line indicate significant difference (p < 0.05).

Components such as ash, crude fat and crude fiber were not affected by the method of drying. With respect to protein content, it was lower in the convection-dried leaves, indicating a deterioration of this component in the faster drying process. The NFE was lower in the shade-dried leaves, reflecting the non-deterioration of the protein. The results were similar to those reported by Abou-Arab et al. (2010), who reported values of 5.37% for moisture, 11.41% for protein, 3.73% for crude fat, 15.52% for crude fiber, 7.41% for ash and 61.93% for carbohydrate, in leaves of *S. rebaudiana* dehydrated by shade drying. *S. rebaudiana* leaves have high amounts of crude protein, crude fiber and carbohydrates. Considering the primary nutrient composition of the *S. rebaudiana* leaves, this could be used not only as a sweetener but also as a source of nutrients.

Total Dietary Fiber (TDF), Insoluble Dietary Fiber (IDF), and Soluble Dietary Fiber (SDF)

The dietary fiber is derived from the cell walls and middle lamellas that make up the tissues of plants; these materials are not absorbed in the body since it lacks the enzyme to hydrolyze, but studies have shown that these components can trap carcinogens and other reactive substances such as bile acids, so that their beneficial effects are exerted directly on the small intestine, which not only removes harmful substances but also alters the microenvironment of the colon (bacterial flora composition and pH bile acids) (Sankhala et al., 2005). Values obtained in the characterization of total, insoluble and soluble dietary fiber leaves of *S. rebaudiana*, dehydrated by both convective and shade drying treatments, are presented in Figure 3.

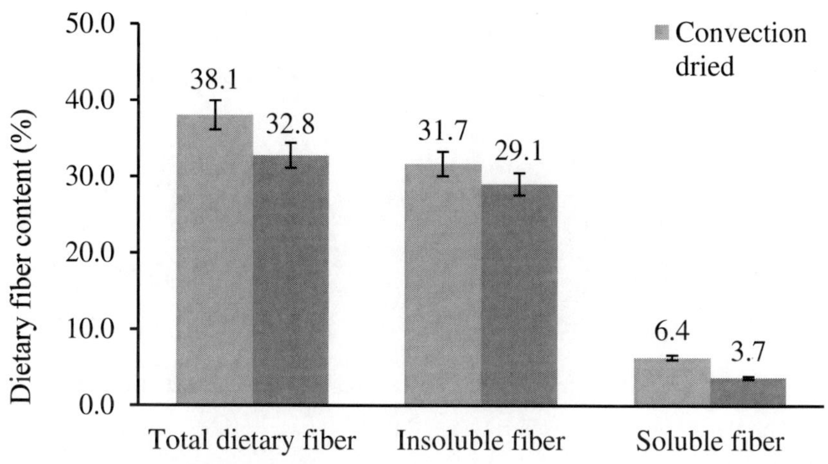

Figure 3. Dietary fiber composition of Stevia rebaudiana Bertoni var. Morita II leaves treated by convective drying and shade drying.

The leaves of *S. rebaudiana* showed a significant TDF content without detecting differences between the drying methods. IDF represented 83.29% of total dietary fiber in the leaves dried by convection, and 88.70% in the shade dried leaves, indicating a higher content in the convection-dried leaves (31.69 ± 0.71 g/100 g leaves) than in the shade-dried leaves (29.07 ± 1.40 g/100 g of leaves). Regarding SDF, it was found that the values were statistically similar for both types of drying: 16.71% for convection-dried leaves and 11.30% for shade-dried leaves respectively. Dietary fiber composition and amounts found in the *S. rebaudiana* leaves make it a potential fiber source for food, due to the beneficial effects of insoluble and soluble components in human health.

Neutral Detergent Fiber (NDF), Acid Detergent Fiber (ADF), and Acid Detergent Lignin (ADL)

Structural carbohydrates, such as cellulose and hemicellulose components, and non-structural polysaccharides and lignin, are included in the NDF. The ADF is comprised of cellulose and lignin, since treatment with the acidic reagent detergent makes the hemicellulose solubilized in its entirety, whereas the ADL – because it is an acid treatment, and virtually leaves only the lignin molecule – is resistant to acid digestions (Van Soest et al., 1991). Content values of NDF, ADF and ADL in the *S. rebaudiana* leaves in both drying treatments are presented in Figure 4.

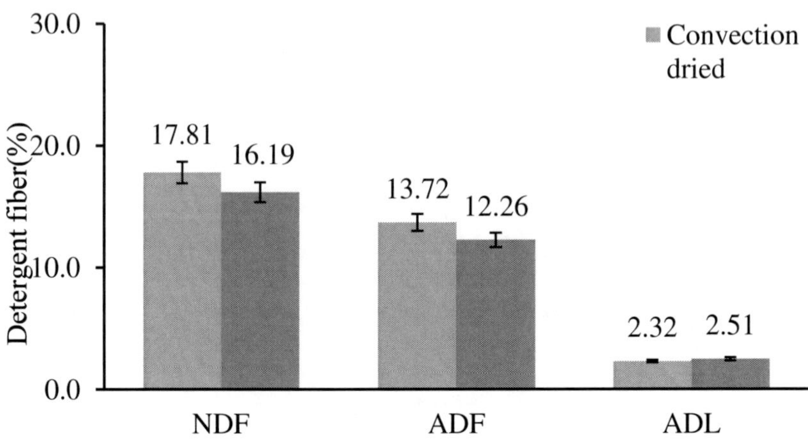

Figure 4. Neutral detergent fiber (NDF), acid detergent fiber (ADF) and acid detergent lignin (ADL) composition of Stevia rebaudiana Bertoni var. Morita II leaves treated by convective drying and shade drying.

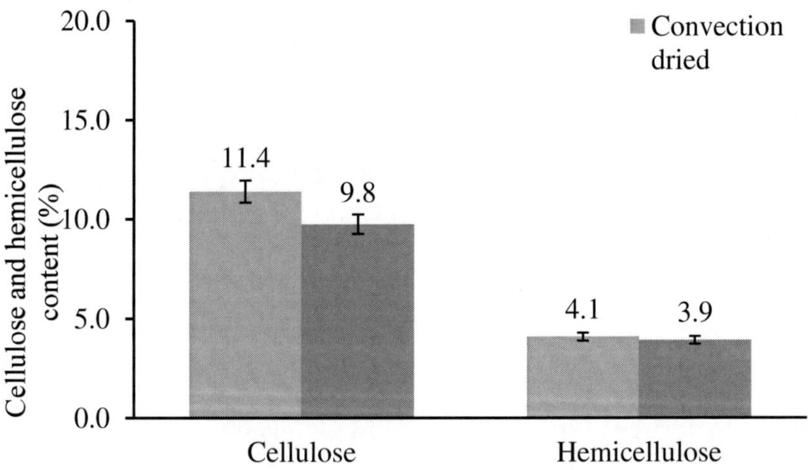

Figure 5. Cellulose and hemicellulose composition of Stevia rebaudiana Bertoni var. Morita II leaves treated by convective drying and shade drying.

The amounts of hemicellulose and cellulose, obtained from leaves after drying, were similar for both drying methods. Cellulose is the principal structural carbohydrate of plants, and an important content found in the leaves of *S. rebaudiana*. This represents a potential use in food applications since it is reported to have poor solubility, it is resistant to digestion in the human gastrointestinal system. These components help to increase the volume of the fecal bolus, promote the peristalsis in the intestine and increase fecal volume (Sánchez-Muñiz, 2012).

Determination of Chlorophylls (A and B) and Carotenoids Contents

Extracts obtained from both dry treatments were analyzed for chlorophyll (a and b), carotenoids and total pigments content (Figure 1). It was found that convection and shade-dried extracts did not differ in their contents of chlorophyll a and b, carotenoids and total pigments.

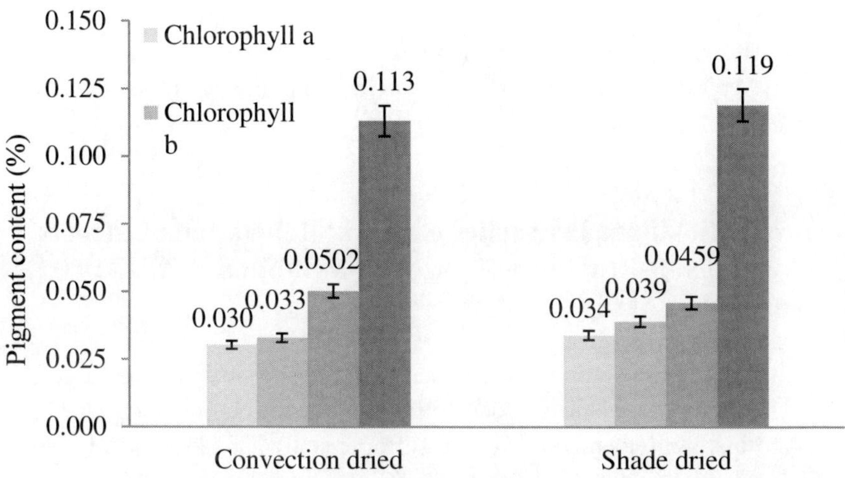

Figure 6. Chlorophyll (a and b), carotenoids and total pigments composition of Stevia rebaudiana Bertoni var. Morita II leaves treated by convective drying and shade drying.

Abou-Arab et al. (2010) quantified the content of chlorophylls (a and b), carotenoids and total pigments in fresh leaves of Stevia, reporting values of 0.111, 0.066, 0.039 and 0.201% of chlorophylls a, b, carotenoids and total pigments, respectively. The same authors reported reductions of these parameters of 47.4, 41.3, 74.8, and 50.9%, with final values of 0.047, 0.027, 0.0076 and 0.075 g/g of chlorophylls a, b, carotenoids and total pigments, respectively, after sun drying and extraction with methanol-water at room temperature. In the present study, the leaves were dried by convection at 60 °C and shade-dried at ambient conditions, and the leaf extracts were obtained with water at 60 °C. Despite this, the contents of chlorophyll b, carotenoids and total pigments were higher in this study (Figure 1). Although chlorophylls are not considered dietary antioxidants, they are widely distributed among green fruit and vegetables, with chlorophyll a and b derivatives predominating in higher plants, and possessing antimutagenic activity and antioxidant activity by breaking the radical chain reaction caused by autoxidation via a hydrogen donating mechanism (Ferruzzi et al., 2002). Meanwhile, carotenoids often occur along with chlorophylls in the chloroplasts, but are also present in other chromoplasts (Krinsky and Johnson, 2005). There are three possible mechanisms for the reaction of carotenoids with radical species, especially singlet oxygen ($^1O_2$) and peroxyl radicals (ROO•), including electron transfer, hydrogen abstraction, and addition of a radical species (Young and Lowe, 2001). While both drying treatments and extraction processes affect the content of chlorophylls and carotenoids, the remaining contents of both pigments could confer antioxidant activity on the extracts of *Stevia rebaudiana*.

In vitro Antioxidant Properties of Stevia Rebaudiana Extracts 2,2´-Azinobis-(3-Ethylbenzothiazoline-6-Sulfonic Acid, Abts•+) Decolorization Assay

This method uses a diode-array spectrophotometer to measure the loss of color when an antioxidant is added to the blue-green chromophore ABTS·+, 2,2'-azino-bis(3-ethylbenzthiazoline- 6-sulfonic acid). The antioxidant reduces ABTS·+ to ABTS, decolorizing it. ABTS·+ is a stable radical. The *Stevia rebaudiana* extracts efficiently scavenged ABTS radicals generated by the reaction between 2,2'-azinobis (3-ethylbenzothiazolin-6-sulphonic acid) (ABTS) and potassium persulfate, with Trolox Equivalent Antioxidant Capacity (TEAC) values of 417 mM/mg sample (convection dried) and 423 mM/mg sample (shade dried). TEAC assay is an excellent tool for determining the antioxidant activity of hydrogen-donating antioxidants and chain-breaking antioxidants (Siow and Hui, 2013).

Ferric Reducing Power

Reducing power was measured by direct electron donation in the reduction of $Fe^{3+}(CN^-)_6$–$Fe^{2+}(CN^-)_6$. The product was visualized by forming the intense Prussian blue color complex; a higher absorbance value indicates a stronger reducing power of the samples. Figure 2 presents the reductive capabilities of extracts of *Stevia rebaudiana*.

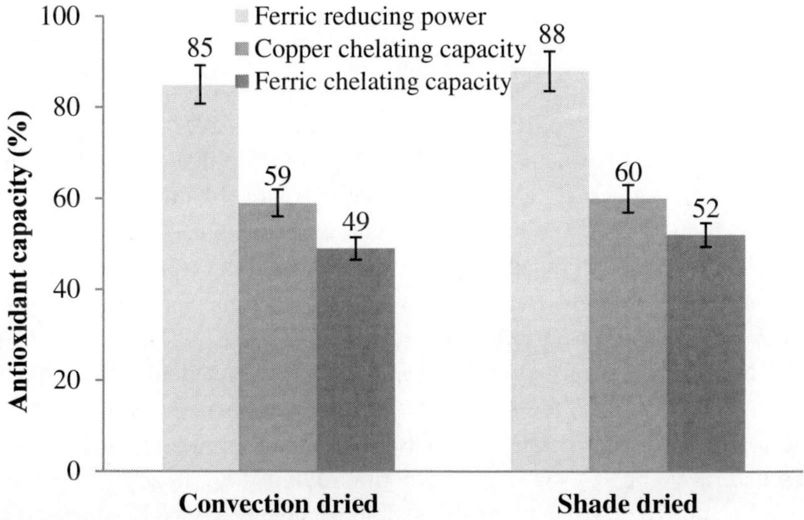

Figure 7. Antioxidant capacities of Stevia rebaudiana Bertoni var. Morita II leaves treated by convective drying and shade drying.

The results of this assay indicate that both extracts are able to donate electrons to reactive radicals, reducing them to more stable and unreactive species.

Chelating of Metal Ions Cu^{2+} and Fe^{2+}

Metal ion chelating activity of an antioxidant molecule prevents oxyradical generation and the consequent oxidative damage. Metal ion chelating capacity plays a significant role in antioxidant mechanisms, since it reduces the concentration of transition metal ions, which are potent catalysts and capable of initiating processes in lipid peroxidation, especially in cellular membranes. Metal ion chelating capacities of the *Stevia rebaudiana* leaf extracts are shown in Figure 2. The main strategy to avoid reactive oxygen species generation associated with redox active metal catalysis involves metal ion chelating. Both *Stevia rebaudiana* extracts interfered with the formation of ferrous and cupric complexes, suggesting that they have chelating activity and can capture metallic ions.

CONCLUSION

The different drying techniques had an effect on the primary nutrients, fiber and antioxidant capacities of *S. rebaudiana* leaves. For preserving primary nutrients, fiber and antioxidant capacities of leaves, both drying treatments were effective. Convective drying is suitable for modern food processing industries, while shade drying will be a low-cost alternative for farmers. The extracts obtained from the leaves of *Stevia rebaudiana* Bertoni were found to be effective radical scavengers, and also possessed a good reducing power and chelating activity. Earlier reports on the antioxidant activity of varieties of *Stevia rebaudiana* Bertoni adapted to cultivation in Mexico are very rare in the literature. The high antioxidant activity of the Mexican variety of *Stevia rebaudiana* Bertoni enhanced the potential interest in this natural source of sweeteners, for improving the efficacy of different products as nutraceutical and pharmacological agents. The consumption of *Stevia rebaudiana* Bertoni may play a role in preventing human diseases in which free radicals are involved, such as cancer, cardiovascular disease and aging. The results of the present study suggest the potential use of the processed (convection dried and shade dried) leaves of *Stevia rebaudiana* var. Morita II as functional foods for their high nutrient and fiber content, and their multiple antioxidant capacities.

REFERENCES

Abou-Arab, A.E., Abou-Arab, A.A., Abu-Salem, M.F. (2010). Physico-chemical assessment of natural sweeteners steviosides produced from *Stevia rebaudiana* Bertoni plant. *African Journal of Food Science, 4*, 269-281.

AOAC. (1998). Official Methods of Analysis, Association of Official Analytical Chemists, Arlington,VA. Secs. 920.39, 923.03, 925.09, 954.01, 962.09.

Atteh, J., Onagbesan, O., Tona, K., Buyse, J., Decuypere, E., Geuns, J. (2011). Potential use of *Stevia rebaudiana* in animal feeds. *Archivos de Zootecnia, 60(229)*, 133-136.

Bhuiyan, M.H.R., Alam, M.M., Islam, N. (2011). The construction and testing of a combined solar and mechanical cabinet dryer. *Journal of Environmental Science and Natural Resources, 4(2)*, 35-40.

Carreras, M.C., Franco, M.C., Peralta, J.G., Poderoso, J.J. (2004). Nitric oxide, complex I, and the modulation of mitochondrial reactive species in biology and disease. *Molecular Aspects of Medicine, 25*, 125-139.

Carter, P. (1971). Spectrophotometric determination of serum iron at the submicrogram level with a new reagent (ferrozine). *Analytical Biochemistry, 40*, 450-458.

Cho, J.Y., Park, S-C., Kim, T-W., Kim, K-S., Song, J-C., Kim, S-K., Lee, H-M., Sung, H-J., Park, H-J., Song, Y-B., Yoo, E-S., Lee, C-H., Rhee, M-H. (2006). Radical scavenging and anti-inflammatory activity of extracts from *Opuntia humifusa Raf. Journal of Pharmacy and Pharmacology, 58*, 113-119.

Ebrahimzadeh M.A., Nabvi, S.F., Nabavi, S.M., Eslami, B. (2010). Antihemolytic and antioxidant activities of *Allium paradoxum. Central European Journal of Biology, 5(3)*, 338-345.

Espitia, C.M., Montoya, B.R., Atencio, S.L. (2009). Yield of *Stevia rebaudiana* Bertoni under three population arrangement in middle sinu. *Revista U.D.C.A Actualidad y Divulgación Científica, 12(1)*, 151-161.

Ferruzzi, M.G., Bohm, V., Courtney, P.D. and Schwartz, S.J. (2002). Antioxidant and antimutagenic activity of dietary chlorophyll derivatives determined by radical scavenging and bacterial reverse mutagenesis assays. *Journal of Food Science, 67*, 2589-2595.

Geuns, J.M.C, Augustijns, P., Mols, R., Buyse, J.G., Driessen, B. (2003). Metabolism of stevioside in pigs and intestinal absorption characteristics of stevioside, rebaudioside A and steviol. *Food Chemistry and Toxicology, 41*, 1599-1607.

Krinsky, N.I., Johnson, E.J. (2005). Carotenoid actions and their relation to health and disease. *Molecular Aspects of Medicine, 26*, 459-516.

Lichtenthaller, H.K. (1987). Chlorophylls and carotenoids, pigments of photosynthetic biomembranes. *Methods in Enzymology, 148*, 350-382.

Mahmoudi, M., Ebrahimzadeh, M. A., Ansaroudi, F., Nabavi, S.F., Nabavi, S.M. (2009). Antidepressant and antioxidant activities of *Artemisia absinthium* L. at flowering stage. *African Journal of Biotechnology, 8*, 7170-7175.

Montgomery, D. (2004). Diseño y análisis de experimentos. México, D.F., México: Limusa-Wiley.

Nehir E.L.S. and Karakaya, S. (2004). Radical scavenging and iron-chelating activities of some greens used as traditional dishes in Mediterranean diet. *International Journal of Food Sciences and Nutrition, 55*, 67-74.

Pinnell, S.R. (2003) Cutaneous photodamage, oxidative stress, and topical antioxidant protection. *Journal of the American Academy of Dermatology, 48*, 1-19.

Prosky, L., Asp, N., Schweizer, T., Devries, S., Furda, I. (1988). Determination of insoluble, soluble and total dietary fiber in food and food products: interlaboratory study. *Journal of the Association of Official Analytical Chemists, 71(5)*, 1017-1023.

Ramírez, J.G., Avilés, B.W., Moguel, O.Y., Góngora, G.S., May, L.C. (2011). Estevia (Stevia rebaudiana, Bertoni), un cultivo con potencial productivo en México. Instituto Nacional de Investigaciones Forestales, Agrícolas y Pecuarias. Centro de Investigación Regional sureste.

Rhee, S.J., Lee, C-Y.J., Kim, M-R., Lee C-H. (2004). Potential antioxidant peptides in rice wine. *Journal of Microbiology and Biotechnology, 14*, 715-721.

Saiga, A., Tanabe, S., Nishimura, T. (2003). Antioxidant activity of peptides obtained from porcine myofibrillar proteins by protease treatment. *Journal of Agricultual and Food Chemistry, 51*, 3661-3667.

Sánchez-Muñiz, F.J. (2012). Dietary fibre and cardiovascular health. *Nutrición Hospitalaria, 27(1)*, 31-45.

Sankhala, A., Sankhala, A.K., Bhatnagar, B., Singh, A. (2005). Nutrient composition of less familiar leaves consumed by the tribals of Udaipur region. *Journal of Food Science and Technology, 42(5)*, 446-48.

Shukla, S., Mehta, A., Bajpai, V.K., Shukla S. (2009). *In vitro* antioxidant activity and total phenolic content of ethanolic leaf extract of *Stevia rebaudiana* Bert. *Food Chemistry and Toxicology, 47(9)*, 2338-2343.

Singh, S., Garg, V., Yadav, D., Beg, M.N., Sharma, N. (2012). In vitro antioxidative and antibacterial activities of various parts of *Stevia rebaudiana* (Bertoni). *International Journal of Pharmacy and Pharmaceutical Sciences, 4(3)*, 468-473.

Siow, L.F. and Hui, Y.W. (2013). Comparison on the antioxidant properties of fresh and convection oven-dried guava (*Psidium guajava* L.). *International Food Research Journal, 20(2)*, 639-644.

Sudha, G., Sangeetha, Priya, M.S., Shree, R.I., Vadivukkarasi, S. (2011). *In vitro* free radical scavenging activity of raw pepino fruit (*Solanum muricatum* A.). *International Journal of Current Pharmaceutical Research, 3,* 137-140.

Van Soest, P.J., Robertson, J.B., Lewis, B.A. (1991). Methods for dietary fiber, neutral detergent fiber, and nonstarch polysacharides in relation to animal nutrition. *Journal of Dairy Science, 74,* 358-397.

Verás, A.O.M., Béttega, R., Freire1, F.B., M. A. S. Barrozo, M.A.S., Freire, J.T. (2012). Drying kinetics, structural characteristics and vitamin C retention of pepper (Capsicum baccatum) during convective and freeze drying. *Brazilian Journal of Chemical Engineering. 29(4),* 741-750.

Young, A.J. and Lowe, G.M. 2001. Antioxidant and prooxidant properties of carotenoids. *Archive of Biochemistry and Biophysics, 385,* 20-27.

Chapter 7

STEVIA REBAUDIANA BERTONI ANTIMICROBIAL POTENTIAL

Clara Miracle Belda-Galbis, María Nieves Criado,
Antonio Martínez and Dolores Rodrigo[*]
[1]Instituto de Agroquímica y Tecnología de Alimentos (IATA-CSIC),
Carrer del Catedràtic Agustín Escardino Benlloch, Paterna,
València, Spain

ABSTRACT

Food preservatives are used for their ability to inhibit or slow down microbial multiplication in order to prolong food shelf life. In response to consumer demand for fresh, healthy foods with no chemical additives, nowadays the use of preservatives of animal, plant or microbial origin is encouraged, especially in combination with non-thermal technologies, to obtain minimally processed products that are safe and stable.

The antimicrobial activity of *Stevia rebaudiana* Bertoni (*Stevia*) and therefore its potential use as a natural preservative have recently been studied but not from a kinetic point of view, although mathematical modelling of microbial behaviour and enzyme activity is very useful for the planning of a hazard analysis critical control point (HACCP) system in the industry. For this reason, the antimicrobial properties of different *Stevia* extracts were quantitatively evaluated at different temperatures, a

---

[*]Corresponding author: lolesra@iata.csic.es.

leaf infusion, a crude extract and a purified extract being the *Stevia* extracts under study.

In view of the results obtained, lowering the temperature or increasing the *Stevia* concentration of the leaf infusion, which proved to be the most effective extract to slow down *Listeria innocua* growth, led to an increase in the lag time and a decrease in the growth rate. The effect of reducing the temperature from 37 or 22 to 10 °C, the effect of increasing the *Stevia* concentration from 0 or 0.5 to 1.5 or 2.5% (w/v) at 37 °C, and the elongation of the lag phase observed in the presence of 1.5 and 2.5% (w/v) of *Stevia* at 22 °C were statistically significant.

These results show that *Stevia* could be useful to extend the shelf life and improve the safety margins of minimally processed foods.

1. INTRODUCTION

Given the increasing demand for nutritious, synthetic additive-free foods, nowadays the use of natural preservative ingredients derived from plants, animals or microorganisms to obtain products that satisfy market requirements is encouraged, keeping in mind that preservatives are necessary to maintain the quality, extend the shelf life and ensure the safety of fresh and processed foodstuffs by inhibiting or controlling enzyme activity and microbial growth, even under refrigeration conditions. Prominent among these natural preservatives are some of the antioxidant and antimicrobial substances present in the essential oils of certain herbs and spices (Darughe et al., 2012), although their commercial application is quite difficult because of their volatile nature and their impact on the organoleptic characteristics of foods.

Stevia rebaudiana (*Stevia*) Bertoni, also known as *Stevia*, sweet leaf, sweet herb of Paraguay, honey leaf and candy leaf (Jayaraman et al., 2008; Madan et al., 2010), is a perennial shrub belonging to the Asteraceae family that grows in tropical and subtropical areas of South America, where extracts of its leaves have been used as a natural sweetener for hundreds of years. Nowadays, *Stevia* is included in the formulation of chocolates, sauces, yoghurts, biscuits, soft drinks and juices, replacing sugar totally or partially without drastically affecting the visual acceptability or physical characteristics of foods (Madan et al., 2010), because *Stevia* contains acaloric steviol glycosides up to 450 times sweeter than sucrose (Yadav and Guleria, 2012).

Various studies have shown that dry *Stevia* leaves contain minerals, vitamins, phenols, flavonoids and other compounds (Tadhani and Subhash, 2006; Lemus-Mondaca et al., 2012) that may be beneficial for consumer health and useful for the food industry as preservative agents that, in any case, add value to the food products containing *Stevia*.

Reducing and controlling the concentration and growth of spoilage and pathogenic microorganisms in foods is fundamental to prevent their deterioration and guarantee their safety. Despite our improved understanding of infection processes, better methods for controlling microorganisms and much stricter regulation of food production, foodborne diseases are still a major cause of morbidity and mortality worldwide (Roller, 2003). The inherent ability of microorganisms to mutate and adapt to environmental stressors represents an ongoing food safety challenge (Doyle, 2006), especially given consumer preference for ready-to-eat, minimally processed foods with no synthetic additives. The incorrect or excessive use of existing antimicrobials can create selective pressure that leads to the emergence of resistant microorganisms. Consequently, there is a clear need for new methods for preserving food using natural preservatives, because it is well known that they are rich sources of biologically active compounds (López et al., 2005).

The antimicrobial activity of different *Stevia* extracts has recently been studied because of its ability to inhibit microbial growth on agar (Sivaram and Mukundan, 2003; Tadhani and Subhash, 2006; Debnath, 2008; Ghosh et al., 2008; Jayaraman et al., 2008; Seema, 2010). So far, its potential to accelerate or promote microbial inactivation and/or slow down microbial growth has not been evaluated from a kinetic point of view, taking into account that the preservative effectiveness of any compound or mix depends on incubation conditions.

Studies of this sort, which include mathematical modelling of bacterial behaviour under a broad range of storage conditions (Ross and McMeekin, 1991; Whiting, 1995; Scott et al., 2005), are very useful, however, for assessment of the benefit/risk associated with the incorporation of *Stevia* in the formulation of new products or in the re-formulation of existing ones. For this purpose, *Listeria innocua* growth in the presence of different *Stevia* extracts was modelled and the antimicrobial potential of *Stevia* was assessed on the basis of lag time duration and maximum growth rate reached in each of the scenarios considered, taking into account that *L. innocua* is a non-pathogenic surrogate for *Listeria monocytogenes* (Char et al., 2010; Jadhav et al., 2013).

2. MATERIAL AND METHODS

2.1. Culture Preparation

A vials stock containing the microorganism studied in stationary phase (*L. innocua*, CECT 910) (6.5 × 10^9 cfu/mL) was generated from a lyophilized pure culture provided by the Spanish Type Culture Collection, following the method described by Saucedo-Reyes et al. (2009).

2.2. Experimental Design

Growth curves were obtained at 37 °C, in Tryptic Soy Broth (TSB; Scharlau Chemie, SA, Spain), in the presence of various extracts, at various concentrations. The extracts tested were the following: (i) a leaf infusion prepared from dried leaves (Anagalide, SA, Spain) and bottled water, (ii) a crude extract (GLYCOSTEVIA®-EP, Anagalide, SA, Spain) with a steviol glycosides content equal to or higher than 20%, and (iii) a purified extract of steviol glycosides (GLYCOSTEVIA®-95, Anagalide, SA, Spain) with a high percentage of rebaudioside A (≥ 80%, the product purity being greater than 95%). The antibacterial activity of the most active product was then evaluated at various concentrations (3 concentrations) and temperatures (37, 22 and 10 °C). With this aim, growth curves were obtained in TSB (Scharlau Chemie, SA, Spain).

For each of the conditions studied, two bottles with 20 mL of broth and the *Stevia* extract at the desired concentration were inoculated, the initial inoculum being ca. 1 × 10^5 cfu/mL. With this aim, frozen vials of the available stock were defrosted at room temperature and diluted in 1‰ (w/v) buffered peptone water (Scharlau Chemie, SA, Spain) to achieve the desired inoculum size at time 0, taking into account the final sample volume and the average cell concentration of the vials in stock. Then samples were taken, diluted and seeded in duplicate at regular intervals until the stationary growth phase was reached, using 1‰ (w/v) buffered peptone water (Scharlau Chemie, SA, Spain) and Tryptic Soy Agar (TSA; Scharlau Chemie, SA, Spain), respectively, for the dilution and seeding of the samples, in order to quantify microorganism growth by viable plate count. All the experiments were performed in triplicate and included the obtainment of growth curves in TSB (Scharlau Chemie, SA, Spain) without *Stevia*.

2.3. Growth Modelling

The growth curves obtained for the various combinations of *Stevia* and temperature were fitted to the modified Gompertz equation (Gibson et al., 1988):

$$\log_{10}(N_t) = A + C \times e^{-e^{-B \times (t-M)}} \tag{1}$$

where N_t represents the number of microorganisms (N) at time t (cfu/mL); A the $\log_{10}$ of the initial count (N_0; $\log_{10}$ (cfu/mL)); C the difference between N_{max} and N_0 ($\log_{10}$ (cfu/mL)); B the relative growth rate when $t = M$ (($\log_{10}$ (cfu/mL))/h); M the elapsed time until the maximum growth rate is reached (h), and e is Euler's number, which has a value approximately equal to 2.718.

This model (Equation (1)) was used in accordance with studies previously carried out by other authors to model the effect of natural compounds on bacterial growth (Guillier et al., 2005; Ferrer et al., 2009; Pina-Pérez et al., 2009). A, B, C and M were used to calculate the kinetic parameters lag time (λ, h) and maximum growth rate (μ_{max}, ($\log_{10}$ (cfu/mL))/h)), employing the following equations (Gibson et al., 1988; McMeekin et al., 1993):

$$\lambda = M - \left(\frac{1}{B}\right) + \frac{\log_{10}(N_0) - A}{\mu_{max}} \tag{2}$$

$$\mu_{max} = \frac{B \times C}{e} \tag{3}$$

Data were fitted using the statistical software Statgraphics® Centurion XV (Statpoint, Inc., USA). The accuracy of the fits was determined by means of the adjusted determination coefficient (*adjusted R^2*) and the mean square error (*MSE*), whose mathematical expressions are as follows:

$$adjusted\ R^2 = \left[1 - \frac{n-1 \times \left(1 - \frac{SSQ_{regression}}{SSQ_{total}}\right)}{n-p}\right] \tag{4}$$

$$MSE = \frac{SSQ_{residual}}{n-p} \tag{5}$$

where n represents the number of observations; p the number of model parameters; and SSQ the sum of squares (Saucedo-Reyes et al., 2009).

3. RESULTS AND DISCUSSION

3.1. Effect of Different *Stevia* Extracts on *Listeria innocua* Growth

The effect of *Stevia* on *L. innocua* growth was evaluated by obtaining growth curves of the microorganism in the presence of different extracts: (i) a leaf infusion, (ii) a crude extract, and (iii) a purified extract, under optimal growth conditions, i.e., in reference medium, at 37 °C (Rowan and Anderson, 1998).

On the basis of the experimental data, a marked elongation of the lag phase and a growth rate decrease can be attributed to the infusion and the crude extract. After 5 h of incubation, differences in microorganism concentration ($\log_{10} (N_5)$) were observed between samples with different extracts of *Stevia*, with the lowest concentration of *L. innocua* being found in samples containing the *Stevia* infusion (Table 1). The curves obtained were fitted to the modified Gompertz model and the antimicrobial effect of the *Stevia* extracts was characterized on the basis of the kinetic parameters λ and μ_{max} (Table 1).

Table 1. Modified Gompertz equation fit results for *Listeria innocua* growth at 37 °C, with/without the addition of different *Stevia* extracts

Sample	$\log_{10} (N_5)$ ($\log_{10}$ (cfu/mL))	λ (h)	μ_{max} (($\log_{10}$ (cfu/mL))/h)	adjusted R^2	MSE
Without *Stevia*	8.60 ± 0.15^a	0.99 ± 0.10^a	0.64 ± 0.04^a	0.99	0.02
Purified extract	8.35 ± 0.20^a	0.47 ± 0.06^a	0.58 ± 0.05^a	0.99	0.02
Crude extract	6.58 ± 0.03^b	2.58 ± 0.12^b	0.31 ± 0.04^b	0.99	0.01
Leaf infusion	6.05 ± 0.05^c	10.89 ± 0.28^c	0.25 ± 0.01^c	0.99	0.02

[a-c] Different lowercase letters reflect significant differences between the λ and μ_{max} values obtained in the presence of different *Stevia* extracts.

No differences were detected between the controls and samples containing the purified extract, therefore no bacteriostatic/bactericidal effect against *L.*

innocua can be attributed to it ($p > 0.05$). However, an increase in λ and a decrease in μ_{max} were found for the crude extract and the infusion ($p \leq 0.05$). The crude extract was able to multiply λ by two. With the infusion, the λ value was 10 times higher than the λ value without *Stevia*. μ_{max} was halved in both cases, both in the presence of the crude extract and in the presence of the infusion. Consequently, a statistically significant antimicrobial effect can be attributed to these *Stevia* extracts ($p \leq 0.05$).

Although the use of medicinal plants as sources of natural antimicrobials is booming (Hammer et al., 1999; Radulović et al., 2007), to our knowledge only a few studies have reported the antimicrobial capability of *Stevia* (Sivaram and Mukundan, 2003; Tadhani and Subhash, 2006; Debnath, 2008; Ghosh et al., 2008; Jayaraman et al., 2008; Seema, 2010), and so far no research work has proposed a mathematical model to describe and quantify bacterial growth behaviour in the presence of different *Stevia* extracts.

Differences in antimicrobial capability between the various extracts studied must be due to their different composition. In the present study, the least processed material, the infusion, showed the highest antimicrobial capability against *L. innocua*, followed by the crude extract. The purified extract, which was the most processed of the three, did not seem to affect the growth pattern of *L. innocua* under the conditions studied, probably because the ingredients that contribute antimicrobial capability to *Stevia* are degraded in the process of purification. It is well understood that correlating antimicrobial activity with the phytochemicals present in an extract is a complex task (Ferrer et al., 2005; Nobmann et al., 2009), because the exact composition of any mixture of ingredients determines its antimicrobial capability, and there may even be differences between mixtures that in principle are the same, if they are not obtained at the same time and from the same raw material. For example, it is known that the antimicrobial activity of essential oils depends on the composition of the oil, which in turn depends on the individual plant, the environmental conditions in which it grew, the part from which the oil was obtained and, of course, the extraction process (Karakaya et al., 2011). In general, the process of obtaining purified extracts entails the loss of some potentially antimicrobial ingredients (Nobmann et al., 2009).

Steviol glycoside compounds have been described as diterpenes responsible for the natural sweetness of *Stevia*. Marketed purified extracts from its leaves mainly contain stevioside ($> 80\%$) and rebaudioside A ($> 90\%$). The purified extract under study, which has no apparent antimicrobial activity, contains more than 95% of steviol glycosides, whereas the crude extract possesses a minimum concentration of 20%. The steviol glycoside concentration of *Stevia* leaves has been reported as varying between 4 and 20% (Gardana et al., 2010; Wölwer-Rieck, 2012). It might be thought that the antimicrobial nature of *Stevia* depends on ingredients other than steviol glycosides. The phytochemicals content of *Stevia* leaves has been reported to be rich in flavonoids, alkaloids, chlorophylls, xanthophylls, hydroxycinnamic acids, non-glycoside diterpenes, saponins, sterols and terpenes (Tadhani and Subhash, 2006; Marković et al., 2008). Because of the terpenic chemical structure of steviol glycosides, it cannot be ruled out that they may have antimicrobial properties per se and/or when combined with other compounds (Brandle and Telmer, 2007). Phenolic compounds have been extensively reported to be antimicrobials. According to Muanda et al. (2011), water extract (WE) from *Stevia* leaves had an antimicrobial effect against *Staphylococcus aureus*, *Bacillus subtilis*, *Escherichia coli*, *Pseudomonas aeruginosa* and *Candida albicans*. Muanda et al. (2011) compared the effectiveness of *Stevia* leaf WE, methanol WE (50:50 (v/v)) and essential oil, revealing that the leaf WE showed the highest antimicrobial effect. They determined the compounds identified in each extract, the leaf WE being rich in protocatechin, catechin, rutin, quercetin glycosyl and quercetin dehydrate. Most of the compounds identified by Muanda et al. (2011) are flavonoids. Therefore, according to Choi et al. (2006), the antibacterial activity of *Stevia* extracts could be due mainly to flavonoids, aromatic acids, terpenoids and their ester contents.

In any case, the results obtained show that the *Stevia* crude extract and the infusion studied have antimicrobial properties, so they could be used as calorie-free sweeteners to extend food shelf life.

3.2. EFFECT OF TEMPERATURE AND *STEVIA* CONCENTRATION ON MICROBIAL GROWTH

On the basis of the previous results and because of the antimicrobial effect shown against *L. innocua*, the *Stevia* infusion was selected for a deeper study of *Stevia* concentration and incubation temperature interaction on the antimicrobial capability observed. *Stevia* leaf infusion was prepared at 0, 0.5, 1.5 and 2.5% (w/v) and incubated at 37, 22 and 10 °C to test the impact of the combinations studied on *L. innocua* growth. To quantify *Stevia* concentration and temperature effects, experimental data were fitted to the modified

Gompertz model. Table 2 provides the λ and μ_{max} values obtained in each case as well as the goodness of each fit (*adjusted R^2 and MSE*).

In view of the results obtained, *L. innocua* growth was affected by temperature in the range [37–10 °C], with a significant elongation of λ (10 times) and a significant reduction of μ_{max} (5 times) when the temperature was reduced to 10 °C, in the absence of *Stevia* (control) ($p \leq 0.05$). With the infusion, the result was similar. Temperature reduction from 37 or 22 °C to 10 °C entailed an elongation of λ and a reduction of μ_{max}, with the difference between the values obtained at 10 °C and the values obtained at higher temperatures in the presence of 0.5 and 1.5% (w/v) of *Stevia* being significant, as was also the difference between the values obtained at 10 and at 22 °C in the presence of 2.5% (w/v) of *Stevia*.

Table 2. Modified Gompertz equation fit results for *Listeria innocua* growth at 37, 22 and 10°C, with/without leaf infusion at different concentrations

% *Stevia* (w/v)	Temperature (°C)	λ (h)	μ_{max} ((log$_{10}$ (cfu/mL))/h)	adjusted R^2	MSE
	37	$0.984 \pm 0.175^{A,a}$	$0.625 \pm 0.043^{A,a}$	0.991	0.020
0	22	$2.926 \pm 0.630^{A,d}$	$0.325 \pm 0.007^{B,d}$	0.990	0.026
	10	$10.784 \pm 5.642^{B,g}$	$0.128 \pm 0.036^{C,e}$	0.989	0.026
	37	$1.498 \pm 0.116^{C,a}$	$0.593 \pm 0.033^{D,a}$	0.992	0.018
0.5	22	$3.647 \pm 0.629^{C,d}$	$0.295 \pm 0.021^{E,d}$	0.994	0.012
	10	$11.295 \pm 1.680^{D,g}$	$0.099 \pm 0.017^{F,e}$	0.979	0.036
	37	$3.840 \pm 1.025^{E,b}$	$0.420 \pm 0.085^{G,b}$	0.990	0.022
1.5	22	$5.667 \pm 0.890^{E,e}$	$0.318 \pm 0.012^{G,d}$	0.994	0.014
	10	$12.017 \pm 0.960^{F,g}$	$0.122 \pm 0.019^{H,e}$	0.989	0.029
	37	$9.432 \pm 0.276^{GH,c}$	$0.229 \pm 0.000^{I,c}$	0.993	0.011
2.5	22	$8.193 \pm 0.579^{G,f}$	$0.262 \pm 0.049^{I,d}$	0.990	0.016
	10	$12.365 \pm 1.806^{H,g}$	$0.086 \pm 0.006^{J,e}$	0.988	0.029

[A-J] Different uppercase superscript letters reflect significant differences between the λ and μ_{max} values obtained at different storage temperatures for a given *Stevia* concentration.

[a-g] Different lowercase superscript letters reflect significant differences between the λ and μ_{max} values obtained in the presence of different concentrations of *Stevia* according to temperature.

At 37°C, a significant influence of *Stevia* concentration on λ and μ_{max} values was observed ($p \leq 0.05$). *Stevia* leaf infusion with a concentration of 2.5% (w/v) significantly increased the λ value by up to 9 times at 37 °C, while the μ_{max} values decreased by up to 3 times at 37°C when *Stevia* leaves were prepared at the same concentration, so at optimum growth temperature this natural product was able to slow down the growth of the microorganism studied at concentrations equal to or higher than 1.5% (w/v). At 22 °C, the presence of *Stevia* also increased λ and reduced μ_{max}, the increase in λ observed in the presence of 1.5 and 2.5% (w/v) of *Stevia* being significant. At 10 °C, however, no statistically significant differences were observed between the control samples and the ones containing *Stevia* ($p > 0.05$).

This implies that the antimicrobial effect of the infusion declines as the temperature decreases, or else that the bacterium is more resistant to its effects at low temperatures. It is well known that microorganisms which survive or adapt to a given stress often gain resistance to others (Wesche et al., 2009). When bacteria are grown at low temperatures, they modify their membrane composition to increase their cold tolerance; these changes could also increase *Stevia* resistance (Veldhuizen et al., 2007; Rattanachaikunsopon and Phumkhachorn, 2010). In any case, it seems that *Stevia* could slow down the growth of *L. monocytogenes* if a cold chain failure occurs or if contaminated foods are kept incorrectly.

CONCLUSION

In view of the potential industrial applicability of *Stevia* in the search for natural, healthy sweetness sources, the antimicrobial description of different *Stevia* products has become an important subject for study.

The mathematical modelling of *Stevia* antibacterial capability under different conditions is a first step on the industrial pathway to future incorporation of *Stevia* not only as a sweetener but also as a preservative. Therefore, the current use of this natural product to sweeten foodstuffs and beverages could be reassessed in the light of the additional preservative properties shown during the shelf life of food products if a cold chain failure occurs or if foods are kept incorrectly, especially against pathogens such as *L. monocytogenes*.

ACKNOWLEDGMENTS

The authors thank the Ministry of Economy and Competitiveness for its support through project number AGL2010-22206-C02-01-ALI and the company Anagalide, SA (Spain) for providing them with dried *Stevia* leaves. Clara Miracle Belda-Galbis and María Nieves Criado are especially grateful to the CSIC for providing them with a JAE-Predoc and a JAE-Tec grant, respectively.

REFERENCES

Brandle, J. E. & Telmer, P. G. (2007). Steviol glycoside biosynthesis. *Phytochemistry, 68,* 1855-1863.

Char, C. D., Guerrero, S. N. & Alzamora, S. M. (2010). Mild thermal process combined with vanillin plus citral to help shorten the inactivation time for *Listeria innocua* in orange juice. *Food and Bioprocess Technology, 3,* 752-761.

Choi, Y. M., Noh, D. O., Cho, S. Y., Suh, H. J., Kim, K. M. & Kim, J. M. (2006). Antioxidant and antimicrobial activities of propolis from several regions of Korea. *LWT-Food Science and Technology, 39,* 756-761.

Darughe, F., Barzegar, M. & Sahari, M. A. (2012). Antioxidant and antifungal activity of coriander (*Coriandrum sativum* L.) essential oil in cake. *International Food Research Journal, 19,* 1253-1260.

Debnath, M. (2008). Clonal propagation and antimicrobial activity of an endemic medicinal plant *Stevia rebaudiana. Journal of Medicinal Plants Research, 2,* 45-51.

Doyle, M. P. (2006). Dealing with antimicrobial resistance. *Food Technology, 60,* 22-29.

Ferrer, C., Ramón, D., Muguerza, B., Marco, A. & Martínez, A. (2009). Effect of olive powder on the growth and inhibition of *Bacillus cereus. Foodborne Pathogens and Disease, 6,* 33-37.

Ferrer, M., Soliveri, J., Plou, F. J., López-Cortés, N., Reyes-Duarte, D., Christensen, M., Copa-Patino, J. L. & Ballesteros, A. (2005). Synthesis of sugar esters in solvent mixtures by lipases from *Thermomyces lanuginosus* and *Candida antarctica* B, and their antimicrobial properties. *Enzyme and Microbial Technology, 36,* 391-398.

Gardana, C., Scaglianti, M. & Simonetti, P. (2010). Evaluation of steviol and its glycosides in *Stevia rebaudiana* leaves and commercial sweetener by ultra-high-performance liquid chromatography-mass spectrometry. *Journal of Chromatography A, 1217,* 1463-1470.

Ghosh, S., Subudhi, E. & Nayak, S. (2008). Antimicrobial assay of *Stevia rebaudiana* Bertoni leaf extracts against 10 pathogens. *International Journal of Integrative Biology, 2,* 27-31.

Gibson, A. M., Bratchell, N. & Roberts, T. A. (1988). Predicting microbial growth: Growth responses of salmonellae in a laboratory medium as affected by pH, sodium chloride and storage temperature. *International Journal of Food Microbiology, 6,* 155-178.

Guillier, L., Pardon, P. & Augustin, J. C. (2005). Influence of stress on individual lag time distributions of *Listeria monocytogenes. Applied and Environmental Microbiology, 71,* 2940-2948.

Hammer, K. A., Carson, C. F. & Riley, T. V. (1999). Antimicrobial activity of essential oils and other plant extracts. *Journal of Applied Microbiology, 86,* 985-990.

Jadhav, S., Shah, R., Bhave, M. & Palombo, E. A. (2013). Inhibitory activity of yarrow essential oil on *Listeria* planktonic cells and biofilms. *Food Control, 29,* 125-130.

Jayaraman, S., Manoharan, M. S. & Illanchezian, S. (2008). In vitro antimicrobial and antitumor activities of *Stevia rebaudiana* (Asteraceae) leaf extracts. *Tropical Journal of Pharmaceutical Research, 7,* 1143-1149.

Karakaya, S., El, S. N., Karagözlü, N. & Sahin, S. (2011). Antioxidant and antimicrobial activities of essential oils obtained from oregano (*Origanum vulgare* ssp. *hirtum*) by using different extraction methods. *Journal of Medicinal Food, 14,* 645-652.

Lemus-Mondaca, R., Vega-Gálvez, A., Zura-Bravo, L. & Ah-Hen, K. (2012). *Stevia rebaudiana* Bertoni, source of a high-potency natural sweetener: A comprehensive review on the biochemical, nutritional and functional aspects. *Food Chemistry, 132,* 1121-1132.

López, P., Sánchez, C., Batlle, R. & Nerín, C. (2005). Solid- and vapor-phase antimicrobial activities of six essential oils: Susceptibility of selected foodborne bacterial and fungal strains. *Journal of Agricultural and Food Chemistry, 53,* 6939-6946.

Madan, S., Ahmad, S., Singh, G. N., Kohli, K., Kumar, Y., Singh, R. & Garg, M. (2010). *Stevia rebaudiana* (Bert.) Bertoni - A Review. *Indian Journal of Natural Products and Resources, 1,* 267-286.

Marković, I. S., Đarmati, Z. A. & Abramović, B. F. (2008). Chemical composition of leaf extracts of *Stevia rebaudiana* Bertoni grown experimentally in Vojvodina. *Journal of the Serbian Chemical Society, 73,* 283-297.

McMeekin, T. A., Olley, J. N., Ross, T. & Ratkowsky, D. A. (1993). Predictive microbiology - Theory and application. Somerset (UK): Research Studies Press Ltd.

Muanda, F. N., Soulimani, R., Diop, B. & Dicko, A. (2011). Study on chemical composition and biological activities of essential oil and extracts from *Stevia rebaudiana* Bertoni leaves. *LWT - Food Science and Technology, 44,* 1865-1872.

Nobmann, P., Smith, A., Dunne, J., Henehan, G. & Bourke, P. (2009). The antimicrobial efficacy and structure activity relationship of novel carbohydrate fatty acid derivatives against *Listeria* spp. and food spoilage microorganisms. *International Journal of Food Microbiology, 128,* 440-445.

Pina-Pérez, M. C., Silva-Angulo, A. B., Rodrigo, D. & Martínez-López, A. (2009). Synergistic effect of pulsed electric fields and CocoanOX 12% on the inactivation kinetics of *Bacillus cereus* in a mixed beverage of liquid whole egg and skim milk. *International Journal of Food Microbiology, 130,* 196-204.

Radulović, N., Mišic, M., Aleksić, J., Đoković, D., Palić, R. & Stojanović, G. (2007). Antimicrobial synergism and antagonism of salicylaldehyde in *Filipendula vulgaris* essential oil. *Fitoterapia, 78,* 565-570.

Rattanachaikunsopon, P. & Phumkhachorn, P. (2010). Assessment of factors influencing antimicrobial activity of carvacrol and cymene against *Vibrio cholerae* in food. *Journal of Bioscience and Bioengineering, 110,* 614-619.

Roller, S. (2003). In Roller, S. (Eds.), *Natural antimicrobials for them inimal processing of foods* (pp. 1-10). UK: Woodhead Publishing Limited & CRC Press LLC.

Ross, T. & McMeekin, T. A. (1991). Predictive microbiology. Applications of a square root model. *Food Australia, 43,* 202-207.

Rowan, N. J. & Anderson, J. G. (1998). Effects of above-optimum growth temperature and cell morphology on thermotolerance of *Listeria monocytogenes* cells suspended in bovine milk. *Applied and Environmental Microbiology, 64,* 2065-2071.

Saucedo-Reyes, D., Marco-Celdrán, A., Pina-Pérez, M. C., Rodrigo, D. & Martínez-López, A. (2009). Modeling survival of high hydrostatic pressure treated stationary- and exponential-phase *Listeria innocua* cells. *Innovative Food Science & Emerging Technologies, 10,* 135-141.

Scott, V. N., Swanson, K. M. J., Freier, T. A., Pruett Jr., W. P., Sveum, W. H., Hall, P. A., Smoot, L. A. & Brown, D. G. (2005). Guidelines for conducting *Listeria monocytogenes* challenge testing of foods. *Food Protection Trends, 25,* 818-825.

Seema, T. (2010). *Stevia rebaudiana*: A medicinal and nutraceutical plant and sweet gold for diabetic patients. *International Journal of Pharmacy & Life Sciences, 1,* 451-457.

Sivaram, L. & Mukundan, U. (2003). In vitro culture studies on *Stevia rebaudiana*. *In vitro Cellular and Developmental Biology-Plant, 39,* 520-523.

Tadhani, M. B. & Subhash, R. (2006). In vitro antimicrobial activity of *Stevia rebaudiana* Bertoni leaves. *Tropical Journal of Pharmaceutical Research, 5,* 557-560.

Veldhuizen, E. J. A., Creutzberg, T. O., Burt, S. A. & Haagsman, H. P. (2007). Low temperature and binding to food components inhibit the antibacterial activity of carvacrol against *Listeria monocytogenes* in steak tartare. *Journal of Food Protection, 70,* 2127-2132.

Wesche, A. M., Gurtler, J. B., Marks, B. P. & Ryser, E. T. (2009). Stress, sublethal injury, resuscitation, and virulence of bacterial foodborne pathogens. *Journal of Food Protection, 72,* 1121-1138.

Whiting, R. C. (1995). Microbial modeling in foods. *Critical Reviews in Food Science and Nutrition, 35,* 467-494.

Wölwer-Rieck, U. (2012). The leaves of *Stevia rebaudiana* (Bertoni), their constituents and the analyses thereof: A Review. *Journal of Agricultural and Food Chemistry, 60,* 886-895.

Yadav, S. K. & Guleria, P. (2012). Steviol glycosides from *Stevia*: Biosynthesis pathway review and their application in foods and medicine. *Critical Reviews in Food Science and Nutrition, 52,* 988-998.

In: Stevia rebaudiana   ISBN: 978-1-63463-335-2
Editors: D. Betancur and M. Segura   © 2015 Nova Science Publishers, Inc.

Chapter 8

ANTIMICROBIAL AND ANTI-TUMOR ACTIVITIES OF *STEVIA REBAUDIANA* BERTONI

J. C. Ruiz-Ruiz, [1], ***Y. B. Moguel-Ordoñez,*** [2]
D. L. Cabrera-Amaro [2] ***and M. R. Segura-Campos***

[1]Departamento de Ingeniería Química-Bioquímica,
Instituto Tecnológico de Mérida, Mérida, Yucatán, México
[2]Instituto Nacional de Investigaciones Forestales,
Agrícolas y Pecuarias, Yucatán, México
[3]Facultad de Ingeniería Química, Universidad Autónoma de Yucatán.
Periférico Norte, Mérida, Yucatán, México

ABSTRACT

Over the years, the World Health Organization (WHO) advocated that countries should interact with traditional medicine with a view to identify and exploit aspects that provide safe and effective remedies for ailments of both microbial and non-microbial origins. The medicinal value of plants lies in some chemical substances that produce a definite physiological action on the human body. The most important of these bioactive compounds of plants are alkaloids, flavonoids, tannins, and phenolic compounds. Many plant leaves have antimicrobial principles

---

* Phone: 52 999 946-09-56, Fax. 52 999 946-09-94. E-mail: maira.segura@uady.mx.

such as tannins, essential oils, and other aromatic compounds. In addition, many biological activities and antibacterial effects have been reported for plant tannins and flavonoids. Plants have an almost limitless ability to synthesize aromatic substances, most of which are phenols or their oxygen-substituted derivatives. These compounds protect the plant from microbial infection and deterioration. Some of these phytochemicals can significantly reduce the risk of cancer due to polyphenol antioxidant and anti-inflammatory effects. Some preclinical studies suggest that phytochemicals can prevent colorectal cancer and other cancers. One of the potent members of the Asteraceae family is *Stevia rebaudiana* Bertoni, which has important industrial uses in beverages, energizers, as well as medicinal uses such as low uric-acid treatment and vasodilator cardiotonic, anesthetic, and anti-inflammatory roles. The aim of the present chapter was to review the antimicrobial and anti-tumor activities of *S. rebaudiana* Bertoni.

INTRODUCTION

Medicinal plants constitute an effective source of both traditional and modern medicines. Herbal medicine has been shown to have genuine utility, and about 80% of rural populations depend on it as primary healthcare. Over the years, the WHO advocated that countries should interact with traditional medicine with a view to identify and exploit aspects that provide safe and effective remedies. The curative parts of a medicinal plant are not simply its woody stem or leaves but the number of chemical compounds (phytochemicals) produced and used for its own growth and development (Bhattacharjee et al., 2005). The therapeutic value and pharmacological action of a drug is due to the presence of certain chemical constituents such as carbohydrates, derivatives of carbohydrates, gums, mucilages, pectins, various forms of glycosides, tannins, phenolic compounds, lipids, fixed and volatile oils, resins, various kinds of alkaloids, and others. These phytochemicals are of immense importance to humankind. Phytochemical investigation of plants is an interesting area of research, leading to the isolation of several new compounds. Knowledge of chemical constituents of plants is desirable not only for the discovery of therapeutic agents but also because such information may be of value in discovering new sources such as tannins, oils, gums, precursors for the synthesis of complex chemical substances, etc. In addition, knowledge of the chemical constituents of plants would be valuable in discovering the actual value of folkloric remedies (Preethi et al., 2011).

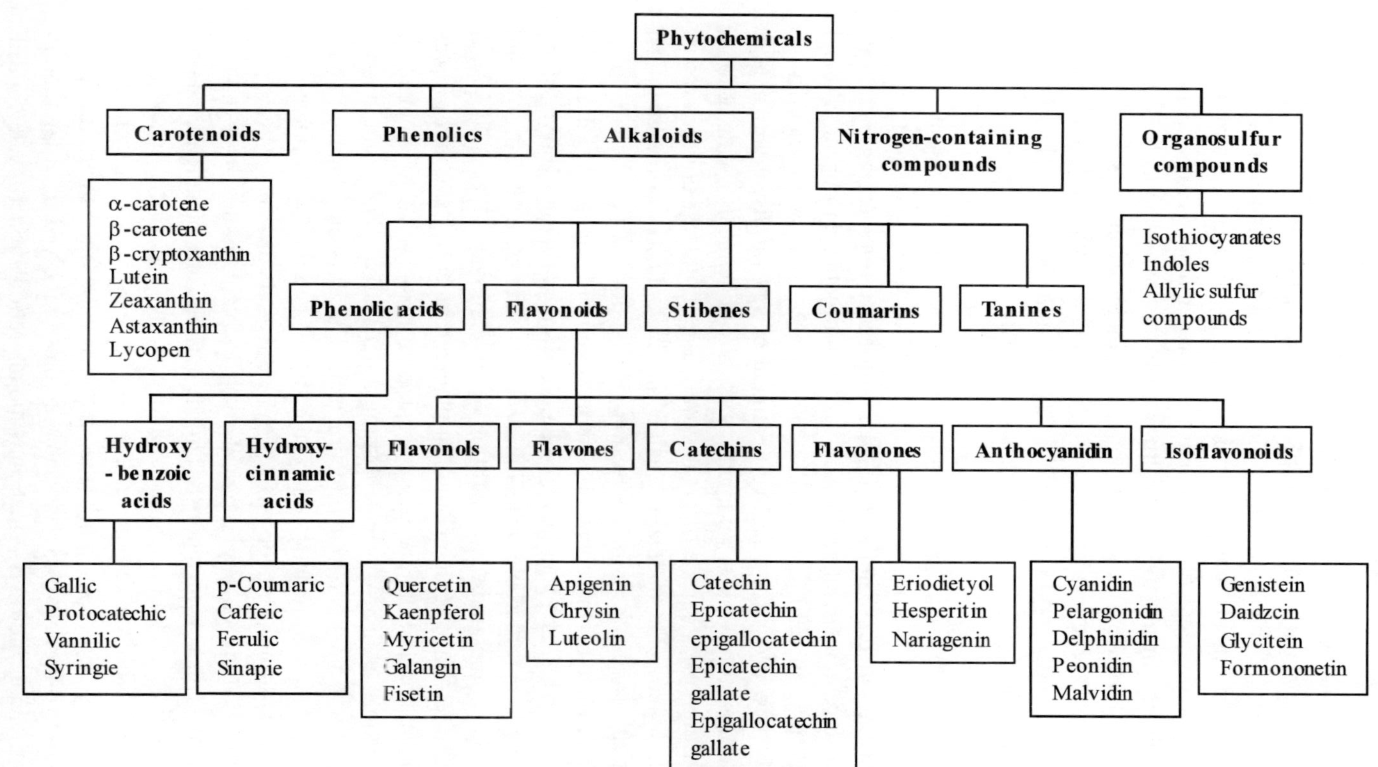

Figure 1. Classification of dietary phytochemicals.

Phytochemicals are defined as bioactive, non-nutrient plant compounds in fruits, vegetables, grains, and other plant foods that have been linked to reducing the risk of major chronic diseases. It is estimated that >5000 individual phytochemicals have been identified in fruits, vegetables, and grains, but a large percentage still remains unknown and needs to be identified before we can fully understand the health benefits of phytochemicals in whole foods (Liu, 2003). However, more and more convincing evidence suggests that the benefits of phytochemicals in fruits and vegetables may be even greater than is currently understood (Liu, 2004). Phytochemicals can be classified as carotenoids, phenolics, alkaloids, nitrogen-containing compounds, and organo-sulfur compounds (Figure 1).

There is a continuous and urgent need to discover new antimicrobial compounds with diverse chemical structures and novel mechanisms of action due to an alarming increase in the incidence of new and re-emerging infectious diseases and development of resistance to the antibiotics in current clinical use. The screening of plant extracts has been of great interest to scientists in the search for new drugs for effective treatment of several diseases (Rojas et al., 2003). Therefore, plant extracts and phytochemicals with known antimicrobial properties can be of great significance in therapeutic treatments (Rojas et al., 2006). The medicinal value of plants lies in some chemical substances that produce a definite physiological action on the human body. The most important of these bioactive compounds of plants are alkaloids, flavonoids, tannins, and phenolic compounds (Edeoga et al., 2005). Many plant leaves have antimicrobial principles such as tannins, essential oils, and other aromatic compounds. In addition, many biological activities and antibacterial effects have been reported for plant tannins and flavonoids. Plants have an almost limitless ability to synthesize aromatic substances, most of which are phenols or their oxygen-substituted derivatives. These compounds protect the plant from microbial infection and deterioration. Some of these phytochemicals can significantly reduce the risk of cancer due to polyphenol antioxidant and anti-inflammatory effects (Yuvaraj et al., 2010).

Due to the variety of physiological roles in plant tissues in regulating enzymes involved in cell metabolism and in mechanisms of defense against foreign agents (i.e., radiations, viruses, and parasites), phytochemicals have been associated with pleiotropic effects in animal cells. Phytochemicals attracted scientists' interests since the demonstration that their biological targets in mammalian cells were the same as those involved in inflammatory processes and oncogenic transformation, such alterations of cell-cycle control, apoptosis evasion, angiogenesis, and metastases. In addition, a large number of epidemiological studies suggest that a daily intake of phytochemicals can reduce the incidence of several types of cancers (D'Incalci et al., 2005; Russo et al., 2005; Sporn and Suh, 2002; Surh, 2003). Many phytochemicals, including polyphenols, are rapidly degraded and metabolized in the human body. Moreover, genetic variation in pathways affecting absorption, metabolism, and distribution of these natural substances could influence exposure at the tissue level, thus modifying disease risk in individuals (Manach et al., 2009). Phytochemicals can prove their therapeutic efficacy in mono-treatments or in association with classical chemotherapeutic drugs. In the latter case, a double-positive effect can be expected: a) phytochemicals can synergize with cytotoxic drugs, increasing their efficacy and lowering the toxic side effects on normal cells; b) combined treatment can delay resistance onset (Russo et al., 2010).

Stevia rebaudiana Bertoni is a plant of recent worldwide focus as a supplement for sugar. The plant originates from South America (Paraguay and Brazil) and belongs to the family Compositae. The plant is also recognized as a medicinal plant in India with its versatile medicinal applications for the treatment of diabetic conditions (acts as non-caloric), candidacies, high blood pressure, weight loss, and skin toner and for its other various uses in the traditional system of medicine (Preethi et al., 2011). *Stevia rebaudiana* is rich in terpenes and flavonoids. The phytochemicals present in *Stevia* are austrinullin, beta-carotene, dulcoside, nilacin, rebaudioxides, riboflavoin, steviol, stevioside, and thiamin. The leaves of this plant are the main attention for economic and commercial applications due to their extreme sweet nature, particularly with the presence of the main active constituents stevioside and rebaudioside A (Khan et al., 2012).

ANTIMICROBIAL ACTIVITY OF *STEVIA REBAUDIANA*

Stevia rebaudiana Bertoni has the ability to inhibit the growth of certain bacteria, which helps to explain its traditional use in treating wounds, sores, and gum disease. It may also explain why the herb is advocated for anyone who is susceptible to yeast infections or reoccurring streptococcal infections, two conditions that seem to be aggravated by white sugar consumption. The biological activity of *Stevia* compounds has been studied by Tomita et al. (1997). They have studied bactericidal activity of a fermented hot-water extract from *Stevia* toward enterohemorrhagic *Escherichia coli* and other food-borne pathogenic bacteria like *Salmonella typhimurium*, *Bacillus subtilis*, and

Staphylococcus aureus. The medicinal properties are attributed to the primary and secondary metabolites synthesized by the plants (Faizi et al., 2003). The effects of plant extracts on bacteria have been studied by a very large number of researchers in different parts of the world. Much work has been done on ethanolic extracts from plants in India (Parekh et al., 2005). The selection of crude plant extracts for screening programs has the potential of being more successful in initial steps than the screening of pure compounds isolated from natural products (Debnath, 2008).

Preethi et al. (2011) performed a preliminary phytochemical analysis of crude extracts of leaves and flowers of *S. rebaudiana* and observed the presence of various phytochemicals such as alkaloids, flavonoids, phenols, steroids, and tannins (Table 1).

Table 1. Qualitative analysis of phytochemicals such as alkaloids, flavonoids, phenols, steroids, tannins in different plant parts of *S. rebaudiana*

Types of extract	Metabolites present in the material				
	Alkaloids	Flavonoids	Phenols	Steroids	Tannins
Leaf					
Ethanol	+++	++++	++++	++++	++++
Methanol	++	+++	+++	++	++
Ethylacetate	++++	++	++	++	++
Chloroform	++++	++	++	++	++
Hexane	++++	++	++	++++	++
Petroleum ether	++	++	+	++	++
Flower					
Methanol	+++	++	++	+	+
Chloroform	+++	++	++	++++	++
Petroleum ether	+++	++	+	+	+

Alkaloids were present in high concentration in chloroform and hexane leaf extracts, and a moderate amount of flavonoids were present in methanol, chloroform, and petroleum ether flower extracts and in ethanol leaf extracts. Flavonoids were present in high concentrations in ethanol leaf extracts. Higher concentrations of phenols were also present in the ethanol leaf extracts when compared to the other extracts.

Lower amounts of phenols are present in the flower extracts. Higher concentrations of steroids were present in ethanol and hexane leaf extracts and in chloroform flower extracts. Lower amounts of steroids are present in the methanol and petroleum ether extracts of flowers. Ethanol leaf extracts contain high concentrations of tannins when compared with the other extracts. Methanol and petroleum ether flower extracts contain fewer amounts of tannins. The preliminary phytochemical studies were of pronounced importance because the crude drugs possess varied compositions of secondary metabolites. The same authors evaluated the antibacterial activity of the extracts against several microorganisms, including *Pseudomonas fluorescence*, *Proteus vulgaris*, *Bacillus subtilis*, *Staphylococcus aureus*, *Klebsiella pneumonia*, and *Streptococcus pneumonia*. All nine crude extracts showed good antibacterial activity. Some of the extracts, like the petroleum ether leaf and flower extracts, gave very low minimum inhibitory concentration (MIC) values. Petroleum ether flower extract gave the lowest MIC values (0.390–1.562 µg/mL) against all the bacterial isolates tested. The lowest MIC value (0.390 µg/mL) was recorded against *B. subtilis*, *Pseudomonas fluorescence*, and *S. pneumoniae*. Methanol flower extracts showed MIC values in the range of 0.781–6.25 µg/mL, whereas leaf methanol extract showed 3.125 and 12.5 µg/mL values, respectively, for all organisms. Of all the extracts tested, the lowest MIC values were recorded in all organisms for the flower rather than leaf extracts.

Khan et al. (2012) used solvents such as water, ethanol, methanol, hexane, acetone, chloroform, and ethyl acetate to extract the secondary metabolites from leaves, stems, and roots of *S. rebaudiana* (Figure 2) and evaluated their antimicrobial activity against *Staphylococcus aureus*, *Escherichia coli*, and *Pseudomonas aeruginosa* using tetracycline as a control. All the extracts were effective against the pathogens used. The most effective extracts from leaves were methanol, ethanol, chloroform, and ethyl acetate. Among the extracts of stems, the most effective were methanol, chloroform, acetone, and ethyl acetate. Among the extracts of roots, the most effective were methanol, chloroform, and acetone. According to these authors, the extracts obtained from leaves showed greater antimicrobial activity than stem and root extracts.

These results demonstrate that antimicrobial activity is correlated with the content of secondary metabolites present in different parts of the plant, which varies depending on the part of the plant.

Gosh et al. (2008) evaluated the antimicrobial activity of *S. rebaudiana* leaf extracts (i.e., petroleum ether, cyclohexane, chloroform, water, acetone, and ethanol) against 10 selected pathogenic as well as food-spoilage fungal (*Alternaria solani, Helminthosporium solani, Aspergillus niger,* and *Pinicillun chrysogenum*) and pathogenic bacterial (*Escherichia coli, Bacillus subtilis, Enterococcus fecalis, Proteus mirabilis, Pseudomonas aeruginosa,* and *Staphylococcus aureus*) isolates, using streptomycin and cotrimazole as controls. These authors found that petroleum ether extracts at 250 µg/mL (MIC) inhibit the growth of *E. coli* and *S. aureus* (by the plate dilution method) among bacteria and *P. chrysogenum* among fungi. Among all extracts, petroleum ether exhibited the best antimicrobial potential followed by water, chloroform, cyclohexane, acetone, and ethanol. This shows that extracts of *S. rebaudiana* act on a wide spectrum of microorganisms.

Figure 2. a) Fresh leaf, stem, root of *Stevia rebaudiana*. b) Powdered leaf, stem, root of *Stevia rebaudiana*.

While most of the extracts obtained with organic solvents showed antimicrobial activity, only those that are obtained with water or ethanol are used in a large number of traditional natural products from plants applied in allopathic medicine. According to Brantner and Grein (1994), these types of plant extracts are potential sources of antimicrobial agents, which could be used as therapeutic agents or preservatives in foods. In this sense, Das et al. (2009) evaluated the antimicrobial activity of aqueous, methanolic, and ethanolic extracts of *S. rebaudiana* leaves. All individual extracts showed potential antimicrobial activity compared to standard ampicillin, but the activities were lower than standard. However, aqueous extracts showed a high, statistically significant antimicrobial activity against *B. substilis*, *S. aureus*, *E. coli*, and *S. typhi*. Aqueous extracts of *S. rebaudiana* exhibit high, significant antimicrobial activity not only against bacteria but also against food-spoilage fungi such as *A. solani*, *H. solani*, *A. niger*, and *P. chrysogenum* (Gosh et al., 2008) and fungi and yeast such as *Candida albicans*, *Cryptococcus neoformans*, *Trichophyton mentagrophytes*, and *Epidermophyton species*) (Jayaraman et al., 2008).

Usually, the extracts are obtained at room temperature, but a higher temperature has influence on the antimicrobial activity. Khan et al., (2012) evaluated the antimicrobial activity against *S. aureus*, *E. coli*, and *P. aeruginosa* (tetracycline as a control) of *S. rebaudiana* leaf, steam, and root aqueous and ethanolic extracts obtained at 80 and 70 °C, respectively. For leaf aqueous and ethanolic extracts, the temperature increased the antimicrobial activity, although for stem and root extracts the activity decreased.

The results obtained by various authors indicate that the solubility and concentration of secondary metabolites are responsible for the antimicrobial activity of the different extracts. When a dilution of the plant extract (for determining of the minimum inhibitory concentration) occurs, better susceptibility and inhibition is observed in many cases. This may be due to purity of the extract, in that prior to dilution it was more viscous and unable to permeate and diffuse properly in the medium (well-diffusion assay) but, following dilution, could easily diffuse into the medium. Hence, *S. rebaudiana* can be further subjected to isolation of antimicrobial compounds. Therefore, these compounds could be proven as future potential candidates either as non-antibiotic pharmaceuticals of food preservatives and/or plant micro-biocides after proper toxicity studies in plant and animal models and clinical trials are addressed.

ANTI-TUMOR ACTIVITY OF *STEVIA REBAUDIANA*

Cancer is a group of diseases caused by loss of cell-cycle control. Cancer is associated with abnormal, uncontrolled cell growth (Krishnamurthi, 2007). Cancer is caused by both external factors (tobacco, chemicals, radiation, and infectious organisms) and internal factors (inherited mutations, hormones, immune conditions, and mutations that occur from metabolism). Cancer is a significant worldwide health problem generally due to the lack of widespread and comprehensive early detection methods, the associated poor prognosis of patients diagnosed in later stages of the disease, and its increasing incidence on a global scale. Indeed, the struggle to combat cancer is one of the greatest challenges of humankind (Divisi et al., 2006). To date, several scientific studies focused on the activity of non-nutritional compounds present in the diet, preventing the occurrence of degenerative diseases such as cancer. This heterogeneous class of molecules, generally known as phytochemicals, includes vitamins (carotenoids) and food polyphenols, such as flavonoids, phytoalexins, phenolic acids, indoles, and sulfur-rich compounds (Russo, 2007; Sporn and Suh, 2002; Surh, 2003). More than 10,000 phytochemicals have been described, and among them more than 6,000 compounds are included in the class of flavonoids. They are widely present in plant-derived foods, beverages (fruits, vegetables, and beverages such as tea, wine, beer, and chocolate), and many dietary supplements or herbal remedies (Russo et al., 2010). According to Cragg and Newman (2000), more than 50% of the drugs in clinical trials for anti-cancer properties were isolated from natural sources or are related to them. Several natural products of plant origin have potential value as chemotherapeutic agents. Some of the currently used anti-cancer agents derived from plants are podophyllotoxin, taxol, vincristine, and camptothecin (Pezzuto, 1997). The areas of cancer and infectious diseases have a leading position in utilization of medicinal plants as a source of drug discovery. Among Food and Drug Administration approved anti-cancer and anti-infectious drugs, drugs from natural origin have a share of 60 and 75%, respectively (Newman et al., 2003).

A great number of *in-vitro* and *in-vivo* methods have been developed to measure the efficiency of natural anti-cancer compounds either as pure compounds or as plant extracts. *In-vitro* methods including the tryphan blue dye exclusion assay, LDH (lactic dehydrogenase) assay, MTT (3-[4,5-dimethylthiazol-2-yl]-2,5 diphenyl tetrazolium bromide) assay, XTT (2,3-Bis-(2-Methoxy-4-Nitro-5-Sulfophenyl)-2H-Tetrazolium-5-Carboxanilide) assay, and sulforhodamine B assay are most commonly used for estimating anti-cancer properties of natural products from medicinal plants. Among all *in-vitro* methods, MTT and the sulforhodamine B assay are the most popular for estimating anti-cancer activity (Chanda and Nagani, 2013). However, there is a discrepancy between phytochemical concentrations applied in *in-vitro* studies (usually tens of micromolars) and those found *in vivo* (human and animal sera) after vegetable and fruit ingestion (usually below 1 µM) (Russo, 2007). Nevertheless, this wide group of natural molecules represents a promising class of anti-cancer drugs due to their multiple targets in cancer cells and limited toxic effects on normal cells.

The anti-tumor potential of *Stevia rebaudiana* was evaluated *in vitro* by Jayaraman et al. (2008). These authors obtained leaf extracts with organic solvents (i.e., ethyl acetate, acetone, and chloroform) and water. These authors performed an anti-tumor assay on human laryngeal epithelioma cells (HEp2). The aqueous extract of *S. rebaudiana* showed no pronounced anti-tumor activity, but the acetone and ethyl acetate extracts were more cytotoxic to HEp2 cells. Acetone extracts showed the highest cytotoxic activity, followed by ethyl acetate and chloroform extracts. An MTT (3-[4,5-dimethylthiazol-2-yl]-2,5 diphenyl tetrazolium bromide) assay was used to evaluate cytotoxicity based on metabolic reduction of MTT. In treatments with Vero (African green monkey kidney cells) cells, 1:2 and 1:4 dilutions of the acetone extract showed cytotoxicity, but there was no apparent cytotoxicity at 1:8 dilution. Further dilutions also had no toxic effects on Vero cells. The 1:2 and 1:4 dilutions were cytotoxic to HEp2 cells, whereas the 1:8 dilution caused more than 50% cytotoxicity and also cessation of cell growth. Further dilutions had less effect on the viability of the cancerous cells. Thus, the 1:8 dilution of the acetone extract of *S. rebaudiana* is non-toxic to the normal cells and also has both anti-cancer and anti-proliferative activities against the cancerous cells.

In another study, Rajesh et al. (2010) evaluated the anti-cancer activity of an ethanolic extract of leaves of *S. rebaudiana* in rats by induced Erlisch's Ascites carcinoma. The effect of *S. rebaudiana* ethanolic extract on tumor growth by hematological parameters is presented in Table 2. The authors concluded that the ethanolic extract has shown good anti-cancer activity at the dose level of 300 mg/kg/i.p (intraperitoneal injection). The extract decreased the tumor volume, the viable and non-viable tumor cell count, and the bodyweight as well as increased the lifespan in a dose-dependent manner compared to the standard drug (5-fluoro uracil at 20 mg/kg/i.p).

Table 2. Effect of ethanolic extract of *S. rebaudiana* treatment on tumor growth by hematological parameters

Groups	Tumor volume (mL)	Viable tumor cells count x 10^7	Non-viable tumor cells count x 10^7	Increase of life span (%)	Body weight increase (g)
Control	4.9 ± 0.94	10.6 ± 0.31	0.4 ± 0.03	23.2 ± 1.91	39.21 ± 0.3
S. rebaudiana extract (100 mg/kg)	1.6 ± 0.93	3.5 ± 0.05	0.8 ± 0.03	67.4 ± 2.27	31.54 ± 0.9
S. rebaudiana extract (300 mg/kg)	0.8 ± 0.28	2.9 ± 0.08	4.9 ± 0.94	4.9 ± 0.94	4.9 ± 0.94
5-Fluoro Uracil (20 mg/kg)	1.9 ± 0.48	3.1 ± 0.17	0.9 ± 0.01	72.4 ± 1.54	28.49 ± 0.02

The ethanolic extract also influenced other hematological and biochemical parameters, increased the hemoglobin level, red blood cell count, lymphocyte and monocyte counts, glucose level, protein level, and contents of urea and uric acid, and decreased the packed cell volume, neutrophils count, and cholesterol level. These results points to the probable anti-tumor potential of some solvent extracts of *S. rebaudiana* leaves. There is a need for further investigation of this plant in order to identify and isolate its active anti-cancer principles. The results of these studies will also need to be confirmed using *in-vivo* models and clinical trials.

CONCLUSION

This review points to the probable antimicrobial and anti-tumor potentials of some solvent extracts of *Stevia rebaudiana*. Extracts obtained from leaves, stems, flowers, and roots exhibit biological activity. There is a need for further investigation of this plant in order to identify and isolate its active, non-antibiotic pharmaceuticals and anti-cancer principles. The results of the study will also need to be confirmed using *in-vivo* models and clinical trials. The final objective is translating the anti-cancer efficacy of phytochemicals into clinics, where these anti-cancer compounds could be administered in association with well-known drugs currently used in chemotherapy. From a pharmacologic point of view, this strategy presents several advantages. Phytochemicals are functionally pleiotropic: they possess multiple intracellular targets, affecting different cell-signaling processes usually altered in cancer cells with limited toxicity on normal cells. Targeting simultaneously multiple pathways may help to kill cancer cells and slow drug-resistance onset. In addition, if proven, the association of pure or synthetic analogs of phytochemicals with chemotherapy or radiotherapy may take advantage of the synergic effects of the combined protocols, resulting in the possibility to lower doses and, consequently, reduce toxicity.

REFERENCES

Bhattacharjee, I., Ghosh, A., Chandra, G. (2005). Antimicrobial activity of the essential oil of *Cestrum diurnum* L. *African Journal of Biotechnology, 4(4)*, 371-374.

Brantner, A. and Grein, E. (1994). Antibacterial activity of plant extracts used externally in traditional medicine. *Journal of Ethnopharmacology, 44*, 35-40.

Chanda, S. and Nagani, K. (2013). *In vitro* and *in vivo* methods for anticancer activity evaluation and some Indian medicinal plants possessing anticancer properties: An overview. *Journal of Pharmacognosy and Phytochemistry, 2(2)*, 140-152.

Cragg, G.M., Newman, D.J. (2000). Antineoplastic agents from natural sources: achievements and future directions. *Expet Opin Investig Drugs, 9*, 1-15.

Das, K., Dang, R., Gupta, N. (2009). Comparative antimicrobial potential of different extracts of leaves of *Stevia rebaudiana* Bert. *International Journal of Natural and Engineering Sciences, 3(1)*, 65-68.

Debnath, M. (2008). Clonal propagation and antimicrobial activity of an endemic medicinal plant *Stevia rebaudiana*. *Journal of Medicinal Plants Research, 2(2)*, 045-051.

D'Incalci, M., Steward, W.P., Gescher, A.J. (2005). Use of cancer chemopreventive phytochemicals as antineoplastic agents. *Lancet Oncology, 6,* 899-904.

Divisi, D., Di, T.S., Salvemin. S., Garramone, M., Crisci, R. (2006). Diet *and cancer. Acta. Biomedica, 77,* 118-123.

Edeoga, H.O., Okwu, D.E., Mbaebie, B.O. (2005). Phytochemical constituents of some Nigerian medicinal plants. *African Journal of Biotechnology, 4(7),* 685-688.

Faizi, S., Rasool, N., Rashid, M., Khan, R.A., Ahmed, S., Khan, S.A., Ahmad, A., Bibi, N., Ahmed, S.A. (2003). Evaluation of the antimicrobial property of *Polyalthia longifolia* var. pendula: isolation of a lactone as the active antibacterial agent from the ethanol extract of the stem. *Phytotherapy Research, 17(10),* 1177-1181.

Gosh, S., Subudhi, E., Nayak, S. (2008). Antimicrobial assay of *Stevia rebaudiana* Bertoni elaf extracts against 10 pathogens. *International Journal of Integrative Biology, 2(1),* 27-31.

Jayaraman, S., Manoharan, M.S., Illanchezian, S. (2008). *In vitro* antimicrobial and antitumor activities of *Stevia rebaudiana* (Asteraceae) leaf extracts. *Tropical Journal of Pharmaceutical Research, 7(4),* 1143-1149.

Khan, J.A., Mishra, S.K., Vandana, K. (2012). Screening of antibacterial properties of *Stevia rebaudiana. International Journal of Biology, Pharmacy and Allied Sciences, 1(10),* 1517-1523.

Krishnamurthi, K. (2007). Screening of natural products for anticancer and antidiabetic properties. *Health Administrator, 1,* 69-75.

Liu, R.H. (2003). Health benefits of fruits and vegetables are from additive and synergistic combination of phytochemicals. *American Journal of Clinical Nutrition, 78,* 517S-520S.

Liu, R.H. (2004). Potential synergy of phytochemicals in cancer prevention: mechanism of action. *Journal of Nutrition, 134(12),* 3479S-3485S.

Manach, C., Hubert, J., Llorach, R., Scalbert, A. (2009). The complex links between dietary phytochemicals and human health deciphered by metabolomics. *Molecular Nutrition and Food Research, 53,* 1303-1315.

Newman, D.J., Cragg, G.M., Snader, K.M. (2003). Natural products as sources of new drugs over the period 1981-2002. *Journal of Natural Products, 66,* 1022-1037.

Parekh, J., Jadeja, D., Chanda, S. (2005). Efficacy of aqueous and methanol extracts of some medicinal Plants for potential antibacterial activity. *Turkish Journal of Biology, 29,* 203-210.

Pezzuto, J.M. (1997). Plants derived anticancer agents. *Biochemistry and Pharmacology, 53,* 121-133.

Preethi, D., Sridhar, T.M., Josthna, P., Naidu, C.V. (2011). Studies on antibacterial activity, phytochemical analysis of *Stevia rebaudiana* (Bert.). An important calorie free biosweetner. *Journal of Ecobiotechnology, 3(7),* 5-10.

Rajesh, P., Kannan, V.R., Durai, M.T, (2010). Effect of Stevia rebaudiana Bertoni ethanolic extract on anti-cancer activity of Erlisch's Ascites carcinoma induced mice. *Journal of Current Biotica, 3(4),* 549-554.

Rojas, R., Bustamante, B., Bauer, J., Fernández, I., Albán, J., Lock, O. (2003). Antimicrobial activity of selected Peruvian medicinal plants. *Journal of Ethnopharmacol, 88(2-3),* 199-204.

Rojas, J.J., Ochoa, V.J., Ocampo, S.A., Muñoz, J.F. (2006). Screening for antimicrobial activity of ten medicinal plants used in Colombian folkloric medicine: a possible alternative in the treatment of non-nosocomial infections. *BMC Complementary and Alternative Medicine, 6,* 2.

Russo, M., Tedesco, I., Iacomino, G., Palumbo, R., Galano, G., Russo, G.L. (2005). Dietary phytochemicals in chemoprevention of cancer. *Current Medicinal Chemisty-Immunology Endocrine and Metabolic Agents, 5,* 61-72.

Russo, G.L. (2007). Ins and outs of dietary phytochemicals in cancer chemoprevention. *Biochemistry and Pharmacology, 74,* 533-544.

Russo, M., Spagnuolo, C., Tedesco, I., Russo, G.L. (2010). Phytochemicals in cancer prevention and therapy: Truth or dare? *Toxins, 2,* 517-551.

Sporn, M.B. and Suh, N. (2002). Chemoprevention: An essential approach to controlling cancer. *Nature Reviews Cancer, 2,* 537-543.

Surh, Y.J. (2003). Cancer chemoprevention with dietary phytochemicals. *Nature Reviews Cancer 3,* 768-780.

Tomita, T., Sato, N., Arai, T., Shiraishi, H., Sato, M., Takeuchi, M., Kamio, Y. (1997). Bactericidal activity of a fermented hot-water extract from *Stevia rebaudiana* Bertoni and other food-borne pathogenic bacteria. *Microbiology and Immunology, 41(12),* 1005-1009.

Yuvaraj, G.1, Nema, R.K., Thirugnanam, P.E., Nathan, V.S. (2010). *In vitro* antimicrobial and antitumour activities of *Derris brevipes* extracts. *Journal of Chemical and Pharmaceutical Research, 2(5),* 708-714.

In: Stevia rebaudiana
Editors: D. Betancur and M. Segura

ISBN: 978-1-63463-335-2
© 2015 Nova Science Publishers, Inc.

Chapter 9

USE OF *STEVIA REBAUDIANA* EXTRACT AS A SWEETENER OF CHOCOLATES FOR PEOPLE WITH DIABETES

Enrique Barbosa-Martín, Diana Sabido-Cortés, Irma Aranda-Gozález and David Betancur-Ancona[*]
Universidad Autónoma de Yucatán,
Facultad de Ingeniería Química, Mérida, México

ABSTRACT

Chocolate is an appreciated food since ancient times with a positive impact on the health of consumers due to its high content of antioxidants. It is capable to improve mood and works as a cardiovascular protector, among other benefits. In Mexico, there are a high prevalence of obesity and its comorbidities, such as Diabetes Mellitus, which added to a high-calorie food intake pattern, makes these diseases even worse.

The aim of this work was to develop a food product based on cocoa (chocolate) sweetened with *Stevia rebaudiana*, in order to obtain a product that can be consumed by people with diabetes. Based on literature, three formulations were developed where among them; one was selected due to its better taste. The formulation was as follows: cocoa paste 62.4%, cocoa butter 23.9%, *Stevia* extract 9.3%, vanilla 3.8% and soy lecithin 0.6% and the proximate composition was: fat 60%,

---

[*] Email: bancona@uady.mx.

carbohydrates 23%, protein 8%, fiber 4%, moisture 3% and ash 2%. With these data, a nutrimental composition table was made according to the "Norma Oficial Mexicana" NOM-051-SCFI/SSA1-2010. A sensory evaluation was performed with untrained panelists, in order to test two methods of cooling. One sample was cooled at 4 °C, immediately after production and molding, whereas the other was cooled at environment temperature for 24 hours. The results didn't show significant differences between the two methods of cooling (p>0.05) and the acceptation of the samples was above the point of indifference, indicating a general preference for both products.

INTRODUCTION

Cocoa (*Theobroma cacao*) is a tree that grows wild in humid-tropical environments and was domesticated in Central and South America (Waizel, Waizel, Magaña, Campos & San Esteban, 2012) since ancient times. With the fruits of *Theobroma cacao* (cocoa beans), cocoa byproducts are obtained through industrial techniques, such as paste, liqueur, butter, and cocoa powder besides the final products such as chocolate and its derivatives (Quintero & Díaz, 2004).

Chocolate is obtained by roasting, grinding and mixing the cocoa beans with sugar, vanilla and cinnamon (Gutiérrez, 2002). From chocolate, several products are obtained including truffles, pralines, dark or white chocolate bars, chocolate mixed with fruits and nuts, etc. (Quintero & Díaz, 2004). Etymologically the word "chocolate" comes from the Aztec word "xocolatl", meaning "sparkling water." This name was used by the Olmecs, Aztecs and Mayans to identify a bitter, strong flavor and high energy drink, derived from the cocoa (Valenzuela, 2007).

Besides its use as food, scientific evidence reports that eating chocolate could provide a cardioprotective effect by its content of oleic acid, arginine, theobromine, tryptophan, potassium and magnesium. But, fundamentally, this benefit is attributed to its richness in a type of phenolic compounds named flavonoids (Pascual, Valls & Solá, 2009). The mechanisms involved in this cardioprotective effect of cocoa flavonoids fall into five main areas: inhibiting the oxidation of LDL cholesterol and increase HDL (Khan et al., 2012), vasoregulation (Hopper et al., 2008), inhibition of platelet function (Flammer A et al., 2007) and anti-inflammatory effect regulating the immune response (Sies, Schewe, Heiss, & Kelm, 2005; Ramiro-Puig et al., 2007).

On the other hand, *Stevia rebaudiana* is a plant native to Paraguay and grown in subtropical and tropical regions of South and Central America, with sweetening power 300 times that of sucrose (Jarma, Rengifo & Araméndiz H, 2005). The compounds responsible for the sweetening property of the plant are diterpene glycosides named stevioside, rebaudioside A, C, D, and dulcoside A. The main steviol glycosides are stevioside (5-10%), rebaudioside A (2-4%), rebaudioside C (1-2%) and dulcoside A (0.5-1%) (Goyal, Samsher & Goyal, 2009; Jackson et al., 2009) present in the leaves of the plant. These elements are synthesized from acetyl-coenzyme A, in the mevalonic acid pathway (Jarma et al., 2005).

These glycosides have a structure that is not digestible by the human enzymes and, therefore, do not add calories after being consumed. In 2008 the Food and Drug Administration (FDA) in the United States granted the status of "generally recognized as safe" (GRAS) for rebaudioside A to use as a sweetener and in that same year, the Joint Expert Committee on Food Additives (JECFA) and Food Standards Australia New Zealand (FSANZ) established an acceptable daily intake (ADI) of 0-4 mg/kg body weight/day of steviol glycosides (FDA, 2008). Today, *Stevia rebaudiana* glycosides are used in Japan, China, Brazil and it is expanding its market to Europe, Canada and Latin America (Giraldo, Marín & Habeych, 2005).

In Mexico, data from the most recent National Health and Nutrition Examination Survey, (ENSANUT) reported epidemic of overweight and obesity, both in children and adults (Gutiérrez et al., 2012). These conditions become alarming considering that obesity can lead to insulin resistance and diabetes mellitus.

Furthermore, both pathologies, obesity and diabetes are related with the development of cardiovascular disease, such as hypertension, atherosclerosis and dyslipidemia, thereby promoting a progressive deterioration of health (González, Sandoval, Román & Panduro, 2001).

Given that, the new pattern of food consumption in the mexican population is characterized by high-calorie and high sugar food (Sánchez, González, & Pérez, 2010), the aim of this study was to develop a food product based on cocoa (chocolate) sweetened with *Stevia rebaudiana*, in order to obtain a product with the beneficial properties of cocoa but without simple sugars that can be consumed by people with diabetes.

MATERIALS AND METHODS

Materials

The materials used were butter and organic cocoa paste, soy lecithin (Pronat ®), vanilla extract (Pronat ®) and commercial extract of Stevia with 97% of Rebaudioside A (Svetia ®)

Chocolate Formulation and Molding Method

Based on a literature review, a selected formulation (Barbagallo, 2007) was adapted to replace the total sugar content with Stevia maintaining a suitable texture and flavor. Three different formulations (A, B, C) were developed varying the content of cocoa butter, cocoa paste, vanilla extract, soy lecithin, and commercial extract of Stevia (Table 1).

For each formulation, cocoa butter and cocoa paste were melted in separate containers. Subsequently, soy lecithin was added to cocoa butter whereas vanilla and extract of Stevia were added to cocoa paste. Finally, the two blends were incorporated.

Of the three formulations, A was chosen to make the proximal analysis and the sensory evaluation assessment.

To evaluate the cooling and molding method that produces the best texture of chocolate, two batches of formulation A was prepared as described above but one sample was cooled to 4 °C immediately after molding, and the other one was molded and cooled at room temperature for 24 hours.

Table 1. Chocolate formulations sweetened with commercial extract of *S. rebaudiana*

Ingredients (%)	Formulation A	Formulation B	Formulation C
Cocoa paste	62.4	54.79	52.63
Cocoa butter	23.9	30.13	39.47
Stevia extract	9.3	5.47	3.94
Vanilla	3.8	6.84	0
Soy lecithin	0.6	2.73	3.94

Name:______________________________________   Date:______________

Try the chocolate samples that are provided and indicate on the scale, your opinion on them. Mark with an X the line that corresponds to the rating for each sample. After testing each sample please rinse mouth with some water.

Scale	Sample 85	Sample 490
Like very much		
Like much		
Slightly like		
Neither like nor dislike		
Slightly dislike		
Dislike much		
Dislike very much		
COMENTS: ____ ____ 		

Figure 1. Seven point hedonic scale. Where the sample codes represent the two methods of cooling and molding evaluated.

Proximal Analysis

The proximal analysis of formulation A was performed based on the methodologies proposed by the Association of Official Analytical Chemists (AOAC, 1997). Briefly: moisture (method 925.09) was determined by drying the sample in a convection oven at 110 °C for 2 hours; ashes (method 923.03) were quantified by burning the sample at 550 °C in a muffle; protein content (method 920.87) was estimated in a digester by Kjendahl method (N x 6.25); crude fiber (method 962.09) was determined with a sequential acid and base digestion; fat extracted by hexane for one hour was determined (method 920.39); and finally, carbohydrate content was established by difference and expressed as nitrogen-free extract.

Sensory Evaluation

A panel of 70 non-trained judges executed the sensory evaluation of the two modalities of cooling and molding of formulation A. The judges noted the level of pleasure or displeasure through an structured hedonic scale of seven points (Torricella & Pulido, 1989) (Figure 1): three dots indicating higher satisfaction, an intermediate point indicating indifference and three lower points indicating displeasure. The response variable was the level of satisfaction of products and was analyzed based on a statistical design with a completely random distribution. The statistical analysis was performed by analysis of one-way variance where differences were considered significant at p <0.05

RESULTS AND DISCUSION

Among the three formulations developed, formulation A was selected due to its better appearance and flavor and therefore, subsequent analyzes were made only with the formulation A. The proximal composition of chocolate is shown in figure 2, after adjustment of moisture content (3%). With these data and according to Norma Oficial Mexicana NOM-051-SCFI/SSA1-2010 the nutritional label was developed (Figure 3).

In the sensory evaluation, formulation A was assessed considering the two different cooling methods. The result of the analysis of variance was not statistically different (p> 0.05) for both samples with an average of 5.2 (Slightly like) and a mode of 6 (like much) for the two techniques. Of the three formulations, the development team found that the B and C were very greasy to taste and had poor stability at room temperature, whereas formulation A was the most stable due to its lower fat content.

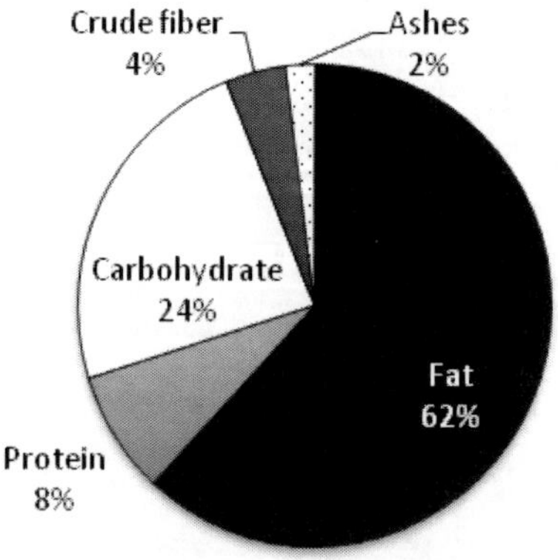

Figure 2. Proximal composition of chocolate (formulation A) sweetened with commercial extract *S. rebaudiana* (% dry basis).

Nutrition Facts

Serving Size 3 bars (30g)
Servings Per Container 1

Amount Per Serving

Calories 150 Calories from Fat 90

	% Daily Value*
Total Fat 10g	**15%**
Saturated Fat 6g	**30%**
Trans Fat 0g	
Cholesterol 0mg	**0%**
Sodium 0mg	**0%**
Total Carbohydrate 12g	**4%**
Dietary Fiber 6g	**24%**
Sugars 0g	
Protein 4g	

Vitamin A 0% • Vitamin C 0%

Calcium 2% • Iron 15%

* Percent Daily Values are based on a 2,000 calorie diet. Your daily values may be higher or lower depending on your calorie needs:

	Calories:	2,000	2,500
Total Fat	Less than	65g	80g
Saturated Fat	Less than	20g	25g
Cholesterol	Less than	300mg	300mg
Sodium	Less than	2,400mg	2,400mg
Total Carbohydrate		300g	375g
Dietary Fiber		25g	30g

Calories per gram:
Fat 9 • Carbohydrate 4 • Protein 4

Figure 3. Nutritional label of chocolate sweetened with commercial extract of *S. rebaudiana*.

**Table 2. Comparison of the nutrient content of chocolate
sweetened with *Stevia* and a regular one**

Content	Chocolate sweetened with *Stevia*	Chocolate sweetened with sugar (Hershey's[®])
Energy content (Kcal)	500	529.26
Fat (g/100 g)	33.33	31.90
Saturated fat (g/100 g)	20	19.51
Carbohydrates (g/100 g)	40	53.65
Sugars (g/100 g)	0	51.21
Fiber (g/100 g)	20	7.31
Protein (g/100 g)	13.3	7.31

Regarding its nutritional composition and according to the guidelines of the Norma Oficial Mexicana NOM-086-SSA1-1994, the product here developed using Stevia as a sweetener can be classified as "sugar free" due to it contains less than 0.5 g of sugars per serving (100 g). The comparison between the compositions of chocolate sweetened with Stevia and a regular one sweetened with sugar is in table 2. It may be noted that the first has 0% sugar, less 5.5% kcal, and 64.5% and 53% more fiber and protein, respectively.

Because simple sugars are rapidly absorbed (Aráuz, 1999), they promote the rise of glucose in blood early after consumption. This effect, on subjects with diabetes who already have problems in their glycemic control could be an aggravating factor associated with the presence of complications (Mendizábal et al., 2010). Thus, both the Norma Oficial Mexicana NOM-015-SSA2 2010 as the American Diabetes Association have recommended to reduce or avoid eating these types of carbohydrates (Bolado, 2002). Therefore, the chocolate here developed may help in the maintenance of glycemic control because it contains no sugar at all. Furthermore, due to its high fiber content (20 g/100 g), this chocolate has the potential to promote metabolic balance in individuals with diabetes; this is based on the capacity that dietary fiber to retain glucose (Escudero & González, 2006) and hydrophobic compounds such as lipids (Villarroel, Acevedo, Yáñez & Biolley, 2003) limiting its absorption and facilitating their excretion. Additionally, this product being a high-protein food would help to maintain a normal body mass in people with diabetes who are prone to muscle wasting (Aráuz, 1999).

It is important to note that besides the sweetness of *Stevia rebaudiana*, some studies have reported that the extracts of S. *rebaudiana* and stevioside have antihyperglycemic effect in animal models (Jeppesen, Gregersen, Alstrup & Hermansen, 2002; Kujur et al., 2010; Lailerd, Saengsirisuwan, Sloniger, Toskulkao & Henriksen, 2004; Misra et al., 2011; Raskovic, Gavrilovic, Jakovljevic & Sabo, 2004) and humans (Curi et al., 1986; Gregersen, Jeppesen, Holst, & Hermansen, 2004), due to their capacity to increase insulin production and inhibit gluconeogenesis. Taking all these data into consideration, the use of a chocolate sweetened with *Stevia* could be extremely beneficial to improve glucose concentration as a result of its proximal composition and *Stevia* content.

CONCLUSION

A chocolate sweetened with *Stevia* extract with no sugar, less calorie, more fiber and protein content was developed, with a level of acceptance above the point of indifference (disregard method of cooling); these features make this product suitable to be consumed by people with diabetes with positive health impact.

ACKNOWLEDGMENTS

This publication forms part of the project "Caracterización quimica de las hojas de *Stevia rebaudiana* y su incorporación en alimentos" supported by Fundación Educación Superior Empresa (FESE).

REFERENCES

AOAC. (1997). Official Methods of analysis. Association of Official Analytical Chemists (Vol. 15th). Washington, D.C. USA: Editorial William Horwitz Editor.

Aráuz A. (1999). Recomendaciones nutricionales en Diabetes Mellitus Tipo 2. *Fármacos, 12*(101-108).

Barbagallo, G. (Producer). (2007, 10/10/12). Chocolate, aspectos técnicos. Chocolates y productos de confiteria, una nueva visión del mundo actual.

Bolado V. (2002). Mitos en el tratamiento nutriológico de la obesidad y la diabetes mellitus. *Nutrición Clínica, 5(4),* 267-271.

Curi, R., Alvarez, M., Bazotte, R. B., Botion, L. M., Godoy, J. L., & Bracht, A. (1986). Effect of *Stevia rebaudiana* on glucose tolerance in normal adult humans. *Brazilian Journal of Medical and Biological Research, 19(6)*, 771-774.

Escudero, E., & González, P. (2006). La fibra dietética. *Nutrición Hospitalaria, 21(2)*, 61-72.

Flammer, A., Hermann, F., Sudano, I., Spieker, L., Hermann, M., Cooper, K., Serafini, M., Lüscher, T. F., Ruschitzka, F., Noll, G. & Corti, R. (2007). Dark chocolate improves coronary vasomotion and reduces platelet reactivity. *Circulation, 116(21)*, 2376-2382.

Giraldo, C., Marín, L., & Habeych, I. (2005). Obtención de Edulcorantes de *Stevia rebaudiana* Bertoni. Revista CENIC Ciencias Biológicas, 36(Especial).

González, M., Sandoval, A., Román, S., & Panduro, A. (2001). Obesidad y Diabetes Mellitus Tipo 2. *Investigación en Salud, 3(1)*, 54-60.

Goyal, S. K., Samsher, & Goyal, R. K. (2009). Stevia (*Stevia rebaudiana*) a bio-sweetener: a review. *Internationa Journal of Food Science and Nutrition, 61(1)*, 1-10.

Gregersen, S., Jeppesen, P. B., Holst, J. J., & Hermansen, K. (2004). Antihyperglycemic effects of stevioside in type 2 diabetic subjects. *Metabolism, 53(1)*, 73-76.

Gutiérrez, B. (2002). Chocolate, Polifenoles y Protección a la Salud. *Acta Farmaceutica Bonaerense, 21(2)*, 149-152.

Gutiérrez, J. P., Rivera, J., Shamah, T., Villalpando, S., Franco, A., Cuevas, L., Romero, M., Hernández, M. (2012). Encuesta Nacional de Salud y Nutrición 2012. Resultados Nacionales. Cuernavaca, México: Instituto Nacional de Salud Pública.

Hershey's. (2013). Hershey's Dark Chocolate. Información nutrimental., from http://www.hersheys.com.mx/productos/chocolates/dark-chocolate

Hopper, L., Kroon, P., Rimm, E., Cohn, J., Le, K., Ryder, J., Hall, W.L., Cassidy, A. (2008). Flavonoids, flavonoid-rich foods, and cardiovascular risk: a meta-analysis of randomized controlled trials. *American Journal of Clinical Nutrition, 88(1)*, 38-50.

Jackson, A. U., Tata, A., Wu, C., Perry, R. H., Haas, G., West, L., & Cooks, R. G. (2009). Direct analysis of *Stevia* leaves for diterpene glycosides by desorption electrospray ionization mass spectrometry. *Analyst, 134(5)*, 867-874.

Jarma, A., Rengifo, T., & Araméndiz, H. (2005). Aspectos fisiológicos de estevia (*Stevia rebaudiana* Bertoni) en el Caribe colombiano: Efecto de la radiación incidente sobre el área foliar y la distribución de biomasa. *Agronomía Colombiana, 23(2)*, 207-216.

JECFA. (2010). Steviol glycosides. In FAO (Ed.), FAO JECFA Monographs (pp. 17-21).

Jeppesen, P. B., Gregersen, S., Alstrup, K. K., & Hermansen, K. (2002). Stevioside induces antihyperglycaemic, insulinotropic and glucagonostatic effects in vivo: studies in the diabetic Goto-Kakizaki (GK) rats. *Phytomedicine, 9(1)*, 9-14.

Khan, N., Monagas, M., Andres, C., Casas, R., Urpí, M., Lamuela, R., & Estruch, R. (2012). Regular consumption of cocoa powder with milk increases HDL cholesterol and reduces oxidized LDL levels in subjects at high-risk of cardiovascular disease. *Nutrition, Metabolism & Cardiovascular Diseases, 22(12)*, 1046-1053.

Kujur, R. S., Singh, V., Ram, M., Yadava, H. N., Singh, K. K., Kumari, S., & Roy, B. K. (2010). Antidiabetic activity and phytochemical screening of crude extract of *Stevia rebaudiana* in alloxan-induced diabetic rats. *Pharmacognosy Research, 2(4)*, 258-263.

Lailerd, N., Saengsirisuwan, V., Sloniger, J. A., Toskulkao, C., & Henriksen, E. J. (2004). Effects of stevioside on glucose transport activity in insulin-sensitive and insulin-resistant rat skeletal muscle. Metabolism, 53(1), 101-107. doi: S0026049503003883 [pii]

Mendizábal, T., Navarro, N., Ramírez, A., Cervera, M., Estrada, E., & Ruiz, I. (2010). Características sociodemográficas y clínicas de pacientes con diabetes tipo 2 y microangiopatías. *Anales de la Facultad de Medicina, 17(1)*.

Misra, H., Soni, M., Silawat, N., Mehta, D., Mehta, B. K., & Jain, D. C. (2011). Antidiabetic activity of medium-polar extract from the leaves of *Stevia rebaudiana* Bert. (Bertoni) on alloxan-induced diabetic rats. *Journal of Pharmacy and Bioallied Sciences, 3(2)*, 242-248.

NOM-051-SCFI/SSA1-2010 (2010). Norma Oficial Mexicana, Especificaciones generales de etiquetado para alimentos y bebidas no alcohólicas preenvasados-Información comercial y sanitaria. Secretaría de Salud, Estados Unidos Mexicanos, México, D.F.

NOM-086-SSA1-1994 (1994). Norma Oficial Mexicana, bienes y servicios. Alimentos y bebidas no alcoholicas con modificaciones en su composicion. Especificaciones nutrimentales. Secretaría de Salud, Estados Unidos Mexicanos, México, D.F.

Pascual, V., Valls, R., & Solá, R. (2009). Cacao y chocolate: ¿un placer cardiosaludable? Clínica e Investigación en Arteriosclerosis, 21(04), 198-209.

Quintero, M., & Díaz, K. (2004). El mercado mundial del cacao. *Agroalimentaria*, 18, 47-59.

Ramiro-Puig, E., Pérez-Cano, F., Ramírez-Santana, C., Castellote, C., Izquierdo-Pulido, M., Permanyer, J., Franch, A. & Castell, M. (2007). Spleen lymphocyte function modulated by a cocoa-enriched diet. *Clinical & Experimental Immunology, 149(3)*, 535-542.

Raskovic, A., Gavrilovic, M., Jakovljevic, V., & Sabo, J. (2004). Glucose concentration in the blood of intact and alloxan-treated mice after pretreatment with commercial preparations of *Stevia rebaudiana* (Bertoni). *European Journal of Drug Metabolism and Pharmacokinet, 29(2)*, 87-90.

Sánchez, V., González, G., & Pérez, M. (2010). El nuevo patrón de consumo de alimentos en la sociedad moderna: Educación y Consumismo. Paper presented at the V Foro de Investigación Educativa, México.

Torricella, M. Z., U; Pulido, A. (1989). Evaluación sensorial aplicada al desarrollo de la calidad en la industria sanitaria (1era edición ed.). Habana, Cuba: Editorial Universitaria.

Valenzuela, A. (2007). El chocolare, un placer saludable. *Revista Chilena de Nutrición, 34(3)*.

Villarroel, M., Acevedo, C., Yáñez, E., & Biolley, E. (2003). Propiedades funcionales de la fibra del musgo Sphagnum magellanicum y su utilización en la formulación de productos de panadería. *Archivos Latinoamericanos de Nutrición, 53(4)*.

Waizel, S., Waizel, J., Magaña, J., Campos, P., & San Esteban, J. (2012). Cacao y chocolate: seducción terapéutica. *Anales Médicos, 57(3)*, 236-245.

INDEX

A

acetone, 99, 100, 101, 137, 138, 141
acetonitrile, 52
acid, 11, 26, 32, 44, 46, 52, 64, 69, 70, 85, 96, 101, 102, 103, 107, 108, 110, 129, 132, 149, 151
acidic, 107
active compound, 25, 26, 28, 30, 33
active site, 86
adaptation, 6, 7
additives, 37, 79, 117, 119
adenosine, 22, 63
adenosine triphosphate, 22
adipose tissue, 62
adults, 18, 20, 21, 149
adverse effects, 67, 72
agar, 119
age, 19, 61, 67, 98
air temperature, 10
alcohols, 50
aldehydes, 50
ALI, 89, 127
alkaloids, 50, 71, 124, 131, 132, 134, 136
alters, 106
Amambay mountain range, vii, 2
amino acid(s), vii, 61, 97
amylase, 99
angiogenesis, 134
ANOVA, 83, 88

antagonism, 129
antibiotic, 6, 139, 143
anti-cancer, 97, 140, 141, 142, 143, 145
anticancer activity, 143
anti-diarrhoeal, viii
anti-hyperglycemic, viii, 20, 67, 68
antihypertensive, viii
anti-inflammatory, viii, 97, 113, 132, 134, 148
antioxidant, 20, 60, 68, 69, 70, 71, 72, 73, 74, 75, 77, 78, 79, 80, 89, 90, 91, 96, 97, 98, 102, 103, 104, 110, 111, 112, 113, 114, 118, 132, 134
antitumor, viii, 91, 128, 144
apoptosis, 134
apples, 90
arabinoside, 50, 51
Argentina, 14, 23
arginine, 148
aromatic compounds, 132, 134
arthritis, 97
artificial sweeteners, vii, 2, 3
ascorbic acid, 69, 78, 85, 91, 92
asparagus, 92
assessment, 112, 119, 150
atherosclerosis, 149

B

Bacillus subtilis, 124, 135, 137, 138

C

D

J

K

L

M

O